Genetic Engineering Fundamentals

Genetic Engineering Fundamentals

An Introduction to Principles and Applications

Karl Kammermeyer
University of Iowa
Iowa City, Iowa

Virginia L. Clark
University of Rochester
Rochester, New York

CRC Press is an imprint of the
Taylor & Francis Group, an **informa** business

First published 1989 by Marcel Dekker, Inc.

Published 2019 by CRC Press
Taylor & Francis Group
6000 Broken Sound Parkway NW, Suite 300
Boca Raton, FL 33487-2742

First issued in paperback 2019

ISBN 13: 978-0-367-45113-4 (pbk)
ISBN 13: 978-0-8247-8069-2 (hbk)

Visit the Taylor & Francis Web site at
http://www.taylorandfrancis.com

and the CRC Press Web site at
http://www.crcpress.com

Library of Congress Cataloging-in-Publication Data

Kammermeyer, Karl
Genetic engineering fundamentals : an introduction to principles and applications / Karl Kammermeyer, Virginia L. Clark.
p. cm.
Bibliography: p.
Includes index.
ISBN 0-8247-8069-8 (alk. paper)
1. Genetic engineering. I. Clark, Virginia L. II. Title.
[DNLM: 1. Genetic Engineering. QH 442 K15g]
QH442.K35 1989
575.1'0724–dc19
DNLM/DLC
for Library of Congress 89-1080
CIP

Preface

My interest in genetic engineering resulted from prior research activities at the University of Iowa Hospitals. This association required consistent perusal of medical and biological journals. Some five years ago I became impressed with the multitude of publications dealing with recombinant techniques. I soon realized that there were promising career opportunities in this area for technologists in the chemical and engineering fields.

This work is intended primarily for these and other "nonexperts" who are nevertheless deeply interested in the subject. Chemical technologists and, in particular, chemical engineers are likely to become more involved in future developments and especially in the problems that arise from the scaling up of laboratory techniques.

While the fundamental concepts have been well elucidated and a multitude of research tasks have been successfully completed, there remains the formidable task of scaling up and commercializing the available technological procedures. This will call for a team approach in which the technologists will have to cooperate closely with the knowledgeable guides from the biological sciences.

Preferred training for the technical representatives would call for some experience in biochemistry and microbiology. As present personnel may not have such training, it is my aim to present the essential principles in a reasonably condensed form. Therefore, the themes of this book are the molecular aspects of the involved structures, their interaction with the component parts of the cell, the role of enzymes and processes of cleaving, and the insertion and production of the recombinant molecules. Finally, a most interesting feature is the status of the creation of these sought-after products.

To ease the transitional problem of acquiring a new vocabulary, a fairly extensive glossary is given. Appendices are also included.

A timely discussion of the engineering aspects was published by L. W. Donnelly (*Genetic Engineering: State-of-the Art*, CEP 78 (11)21-22 (1982)). He also points out the valuable experience that is acquired in biochemical practice.

I owe a great deal of thanks to many colleagues at Iowa. Not only did I get answers to my questions, but I also received unexpected support and wishes of success. I hesitate to enumerate the many individuals who helped me as I would regret omitting any one of them. Professionally, the departmental affiliations ranged through biochemistry, botany, chemistry, microbiology, zoology, and the medical sciences. It was a most pleasing experience to deal with the many knowledgeable scientists at the university.

Karl Kammermeyer

Contents

Introduction

GENERAL VIEW

The connotation of *genetic engineering* is manifold. The term was formulated and used by a multitude of investigators in the field of genetics. It is used to describe the activities of biologists, biochemists, microbiologists, and medical research workers, all the way from initial problem identification to synthetic gene manipulation. Ironically, at the present time, there are very few, if any, engineers as such involved in this development process.

To the chemist and the engineer the procedure for handling a problem involves a number of typical steps: identification and definition of the problem, analysis of its component parts, theoretical and experimental verification, and, finally, synthesis of findings into a practical solution and application. The aspect of application is the significant feature of engineering, as expressed so lucidly by Von Karman when he stated that the scientist discovers what is, the engineer creates what has not been.

An excellent book by Laurence E. Karp is *Genetic Engineering: Threat or Promise?* (1). This text, however, deals almost entirely with human genetics and its implication on inheritance of genetic factors. In this respect it represents the identification and definition of the problem phase together with some aspects of analysis. By far the largest amount of published information in recent years deals with the analysis of genes and their identification. A prominent example is the comprehensive paper by Manfred Eigen et al. (2), "The Origin of Genetic Information." This treatise represents a review paper directed largely to an audience that has a limited knowledge of the field. It covers some of the very basic aspects of genetics, such as molecular composition of the genes and their associated structures, the life processes directed by genes, and the possibilities of gene isolation and manipulation to create new, synthetic identities (which represents the true phase of genetic engineering).

SCOPE

To prepare a fairly condensed presentation of what is a vast and growing area of science is a formidable matter. However, brevity is considered essential, and the chief aim is to present a skeletonized version of what should be the highlights and thus stimulate a deeper interest in the subject.

Genetic engineering in its real sense means the synthetic preparation of composite molecules in which foreign DNA has been inserted into a vector molecule. The resulting product is often called a *chimera*—analogous with the chimera of mythology: a creature with the head of a lion, the body of a goat, and the tail of a serpent. This construction of "artifiicial recombinant" molecules is, of course, gene manipulation. Other terms used to describe the procedure are "molecular cloning" or "gene cloning." The chimeras are generally identical organisms that can be cultured so that they amplify the newly created molecule. The new organism then will function to produce the product coded by the gene insertion.

HISTORY

A world-shaking event, unrecognized at its inception and for many decades thereafter, was the presentation of a paper by Gregor Mendel in 1865. The title of the paper was "Experiments on Plant Hybrids," and it constituted the beginning of the study of genetics. An excellent account of this and the gradual unfolding of this effort is presented in the book *Molecular Genetics* by Gunther S. Stent (3).

While the issuing of two U.S. Patents, nos. 4,237,224 and 4,468,464, to here a comprehensive historical review. Suffice it to say that the real developments and significant scientific breakthroughs are of rather recent vintage. One surprising aspect has been the rapidity of new findings, and, thus, an asymptomatic rate of progress has emerged since about 1970.

The modern events leading up to the present decade have been ably summarized by Charles E. Morris in *Food Engineering*, May 1981 (4), and it is worthwhile to present his tabulation herewith (see figure).

While the issuing of two U.S. Patents, nos. 4,237,244 and 4,486,464, to Cohen et al. falls into the time frame covered by the tabulation, the number of pertinent publications is increasing at an exponential rate. Thus, it is becoming very difficult to keep abreast of developments, and it is imperative that competent computerized literature programs be developed (5).

In order to characterize the status of genetic engineering one has to borrow

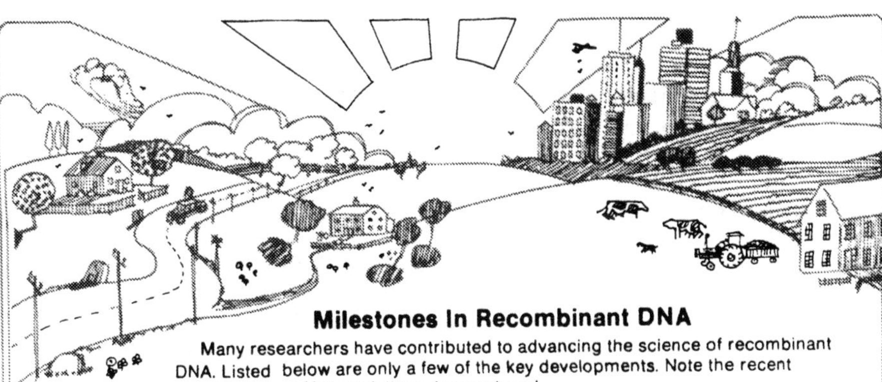

Milestones In Recombinant DNA

Many researchers have contributed to advancing the science of recombinant DNA. Listed below are only a few of the key developments. Note the recent acceleration as NIH guidelines have relaxed.

1944: • Oswald Avery discovers that DNA carries the genetic code

Late 40s • Electron microscope becomes available to scientist at reasonable cost

1952: • Alfred Hershey and Martha Chase apply radioactive labeling to confirm DNA carries the genetic code

1953: • James Watson and Francis Crick discover the double-helix structure of DNA

1955: • Severo Ochoa synthesizes RNA

1956: • Arthur Kornberg synthesizes DNA

1960: • Messenger RNA (mRNA) discovered by researchers at Pasteur Institute in Paris

1971: • Paul Berg constructs first man-made recombinant DNA molecule

1972: • Anada Chakrabarty creates oil-digesting Pseudomonas through cell-fusion of plasmids. Although not recombinant DNA, a major genetic-engineering achievement which ultimately led to landmark 1980 Supreme Court decision.

1973: • Paul Berg demonstrates viral method of recombinant DNA
- Stanley Cohen clones a gene using the plasmid method of recombinant DNA
- O. Wesley McBride and Harvey Ozer demonstrate chromosomes from one species can be made to function in another

1974: • American biologists call for moratorium on recombinant DNA experiments

1975: • Asilomar Conference sets stage for self-regulation of gene-splicing experiments by scientific community

1976: • Recombinant DNA Advisory Committee (RAC) of National Institutes of Health imposes strict guidelines on gene-splicing experiments

1977: • Genentech produces human brain hormone somatostatin through application of recombinant DNA

1978: • Harvard researchers produce rat insulin through recombinant techniques
- Genentech produces human insulin via plasmid method of recombinant DNA
- NIH proposes loosening guidelines as evidence mounts that early concerns were exaggerated
- **Stanford researchers transplant mammalian gene using viral method**
- Genentech licenses Eli Lilly to use its insulin process
- Cornell researchers implant leucine-producing gene into yeast cell via plasmid method

1979: • Genentech produces human growth hormone (HGH) and thymosin alpha-1 through recombinant techniques
- RAC proposes further relaxation of NIH guidelines

1980: • NIH twice publishes progressively relaxed guidelines in Federal Register
- Biogen S.A. and U. of Zurich produce human interferon via recombinant DNA
- Genentech produces human leukocyte interferon, fibroblast interferon and insulin percursor (proinsulin) using recombinant DNA
- Supreme Court rules that life can be patented
- Eli Lilly starts clinical trials of insulin produced by gene-splicing techniques
- Bethesda Research Labs produce amino acid proline by recombinant means
- Cetus applies recombinant DNA to producing enzymes capable of forming ethylene and propylene oxides
- USDA contracts with Genentech to develop vaccine against hoof-and-mouth disease
- Hoffmann-LaRoche, Abbott Labs, G.D. Searle, Pfizer, Schering-Plough, Upjohn and Bristol-Myers all launch recombinant DNA programs for interferon
- Genentech goes public, rocks Wall Street
- U.S. Patent Office awards patent on recombinant DNA plasmid method to Stanford U.

1981 (to date) • Genentech develops bovine growth hormone (BGH) via recombinant DNA techniques
- Interferon produced by Genentech and Hoffmann-La Roche start clinical trials
- Three firms announce plans to market automated gene-synthesis machines, another introduces a computerized protein sequencer
- Genentech produces leukocyte interferon from recombinantly-engineered yeast

Milestones in recombinant DNA. (Reprinted with permission from *Food Technology 35* (7):29 (1981). Copyright © by Institute of Food Technologists.)

terms from the literature of science fiction: "The techniques are fantastic, the results are amazing, and the future is astounding . . ." (6).

REFERENCES

1. Karp, LE. (1976) *Genetic Engineering*, Nelson-Hall, Chicago, IL.
2. Eigen, M. Gardiner W. Schuster P. Winkler-Ostwatitsch, R. (1981). "The Origin of Genetic Information," *Sci. Amer.* 244 (4): 88–118.
3. Stent, GS. (1978). *Molecular Genetics*, 2nd Ed. W.H. Freeman and Co., San Francisco, CA.
4. Morris, CE. (1981). *Food Engineering*, 53 (5): 57–69.
5. Haas, WJ. (1982). "Computing in documentation and scholarly research," *Science*, 215 (4534): 857-861.
6. Asimov, I. (1962). *The Genetic Code*, Signet Books, The New American Library, New York.

1
Basic Components

Considering the complexity of a living organism one is struck by the relatively small number of organic components used in the formation of substances which control the functioning of the organism. For example, the basic control of the "building" of the organism is exercised by two chemicals: deoxyribonucleic acid (DNA) and ribonucleic acid (RNA). These materials consist of polymer chains containing monomer units called nucleotides, which are linked by phosphodiester bridges to form the linear polymer. The nucleotide, in turn, consists of three molecular components: a phosphate radical (Fig. 1), a ribose sugar (Fig. 1), and a nitrogen containing base (Fig. 2).

One may find that the structural formulas of bases are written in different forms by different authors. This is a result of their so-called *tautomeric* behavior. For instance, guanine can be written in its two forms as shown in the bottom line of Figure 2.

NOMENCLATURE OF COMPOUNDS

When the ribose-based complex is separated from the phosphate group the remaining structure is called a nucleoside. So, a phosphonucleoside is a nucleotide. Nucleotides may contain several phosphate residues, but are present in DNA and RNA only as monophosphates. The letter M in the compound name signifies *mono*phosphate, D means di, and T is triphosphate. The conventional nomenclature is shown in Table 1. Because thymine was originally thought to occur only in DNA, the use of the term deoxythymine was considered to be redundant. Both TMP and dTMP are currently used.

Phosphodiester Bridge

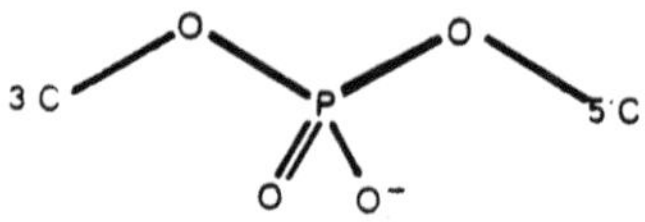

Ribose Sugar

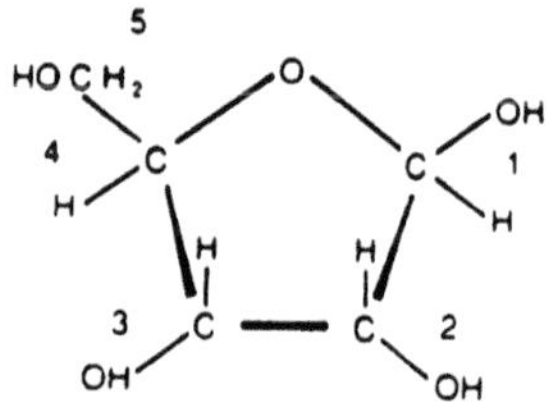

Deoxyribose Sugar

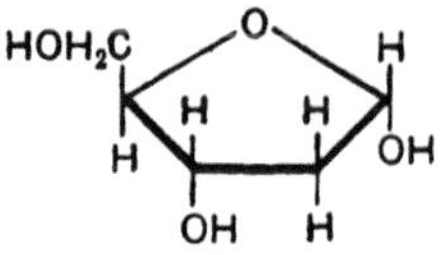

Bases

RNA

Adenine A
Cytosine C
Guanine G
Uracil U

DNA

Adenine A
Cytosine C
Guanine G
Thymine T

FIGURE 1 Formation of nucleotides. Basic components are the phosphodiester bridge, ribose sugars, and bases. Combination rule for bases: Cytosine-C base pairs with Guanine-G; Adenine-A pairs with Uracil-U or Thymine-T (Hydrogen Bonding).

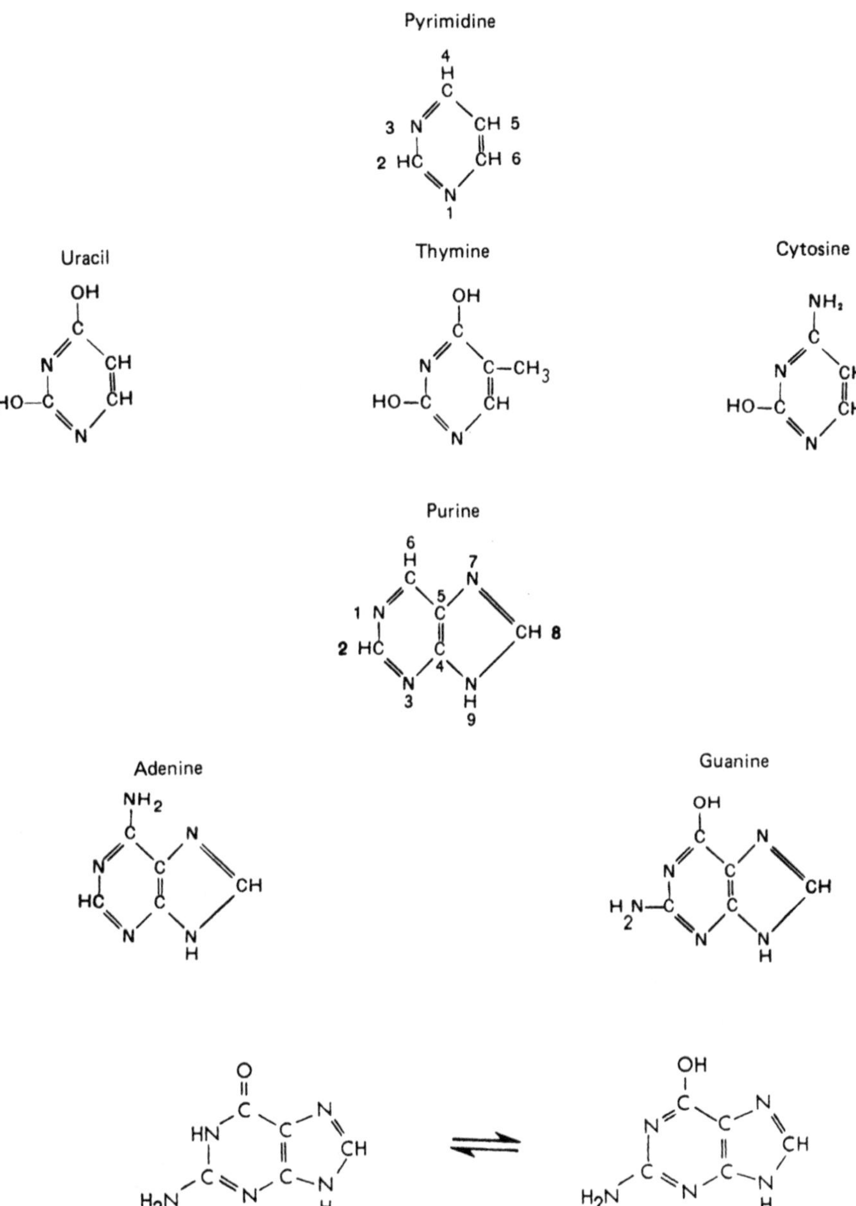

Tautomeric Forms of Bases

FIGURE 2 Basic compounds involved in DNA and RNA formation. The nitrogenous bases, pyrimidines, and purines.

TABLE 1 Conventional Nomenclature of Compounds

Base	Ribonucleoside	Ribonucleotide (5′-monophosphate)
Adenine (A)	Adenosine	Adenylate (AMP)
Guanine (G)	Guanosine	Guanylate (GMP)
Uracil (U)	Uridine	Uridylate (UMP)
Cytosine (C)	Cytidine	Cytidylate (CMP)
	Deoxyribonucleoside	Deoxyribonucleotide (5′-monophosphate)
Deoxyadenine (A)	Deoxyadenosine	Deoxyadenylate (dAMP)
Deoxyguanine (G)	Deoxyguanosine	Deoxyganylate (dGMP)
Deoxythymidine or Thymine (T)	Deoxythimidine or Thymidine	Deoxythymidylate or Thymidylate (dTMP)
Deoxycytosine (C)	Deoxycytidine	Deoxycytidylate (dCMP)

STRUCTURE

The single-stranded DNA polymer molecule associates with a second strand of complementary nucleotide sequence at physiological temperature, pH, and ionic strength. This is less likely to occur with RNA because usually there is no complementary RNA strand; however, it is very common for RNA to form double-stranded regions when complementary sequences occur within an RNA strand. Two DNA strands form a face-to-face configuration resulting in a double helix, and the strands are held together through hydrophobic base-stacking interaction between the adjacent planar bases (~80%) and hydrogen bonds between compatible bases (~20%).

Thus a very strict *combining rule* exists for the hydrogen bonding of pyrimidines and purines (1). A pyrimidine always bonds to a purine and vice versa, and does so in a very specific manner. That is, adenine (A) binds to thymine (T) or uracil (U), and cytosine (C) binds to guanine (G). There are two hydrogen bond locations in the A–T (or U) pairing and three hydrogen bond sites occur in the G–C complex. Note that hydrogen bonds occur only between noncarbon atoms of the base molecules. While this statement describes the opinion most generally held, the situation may change. The March 1, 1982 issue of *Chemical and Engineering News* (2) reports that the first verified hydrogen bond involving carbon atoms, that is –C–H–C, has been found in a ferrocene unit.

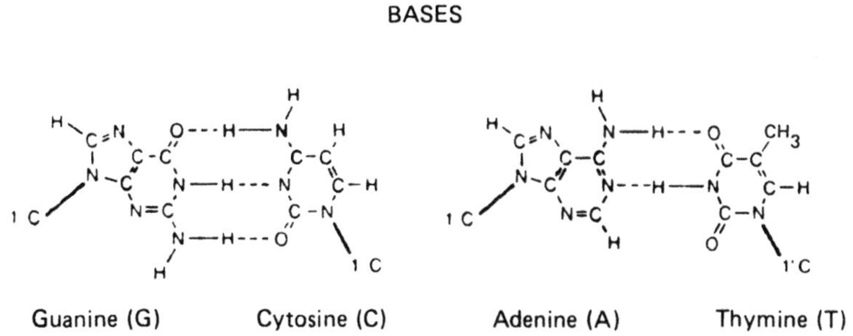

BASES

Guanine (G) Cytosine (C) Adenine (A) Thymine (T)

(a)

DOUBLE-STRANDED DNA

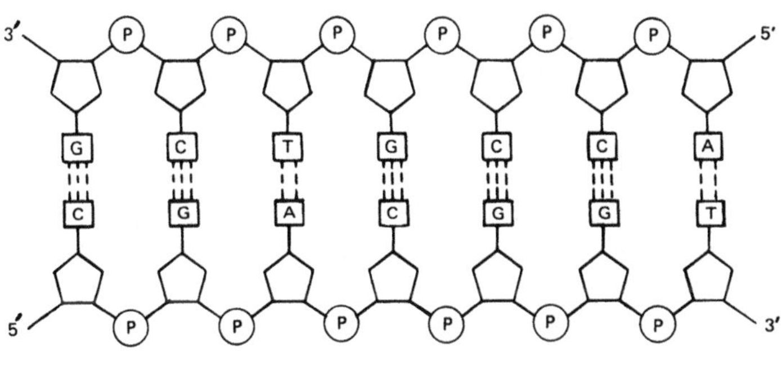

(b)

FIGURE 3 (a) Bonding of bases. Hydrogen bonding of G–C and A–T. (b) Hydrogen bonding in formation of double-stranded DNA.

A simple way to remember the pyrimidine-purine bonding is that alphabetically, the extreme letters A and T (or U) match up (and G–C, accordingly) and so pyrimidine bonds to purine. The A–T (or U) and G–C combinations are called base pairs. The hydrogen bonding and the polymer chain formation are illustrated in Figure 3.

The polymer chains of DNA and most RNAs differ in two aspects only:

DNA	RNA
Deoxyribose	Ribose
Thymine	Uracil
(double-stranded)	(usually single-stranded)

However, transfer and ribosomal RNA also contain a large number of "unusual" bases (3).

Attention should be called to the "numbering" convention. The bases (Fig. 2) are numbered in the usual system for organic compounds. To avoid confusion, the numbering of the ribose sugars (Fig. 1) is by a system of *primes*. This is shown in Figure 3, where the RNA has a 3′ and a 5′ terminal; so the 3′ bond in deoxyribose will combine with the 5′ terminal of the chain.

The 5′ and 3′ positions are of special importance, as they are the reaction sites involved in the mononucleotide addition to the growing DNA polymer chain. It should be noted also that chain growth occurs in the 5′ to 3′ direction by addition of nucleosides and monophosphates. Thus, the 3′–OH radical of the sugar will attach to the 5′-monophosphate end of the nucleoside 5′-triphosphate monomers that combine to form the polymer structure. The DNA synthesis reaction is illustrated in Figure 4 proceeding from the 5′ position, and the growing end of the DNA molecule has an 3′–OH group. The –OH group reacts with the -P atom closest to the ribose sugar (the α-phosphate). The triphosphate radical is attached to the sugar at the 5′ position. The reaction is an esterifcation process and also results in the formation of ionized pyrophosphoric acid $H_4P_2O_7$. The end product is a chain lengthened by one nucleotide, restoring the –OH radical at the 3′ end. The equations are often written in the ionized form, so that an $-O^-$ appears in place of an –OH radical.

The conventional concept of the shape of DNA (and RNA) is that of a double helix (Fig. 5).

SHORTHAND NOTATION

In describing structures or reaction mechanisms of the DNA and RNA chains, it would be cumbersome to always show the respective phosphodiester and ribose structures. The use of the initial letters of bases is already an accepted method for abbreviation and an analogous procedure was adopted for the other nucleotide components. The conventional method, a shorthand notation, is illustrated in Figure 6.

When shorthand notation is used in the text it will appear as follows: pG, meaning a G nucleotide with phosphate on the 5′ end and, correspondingly, Gp,

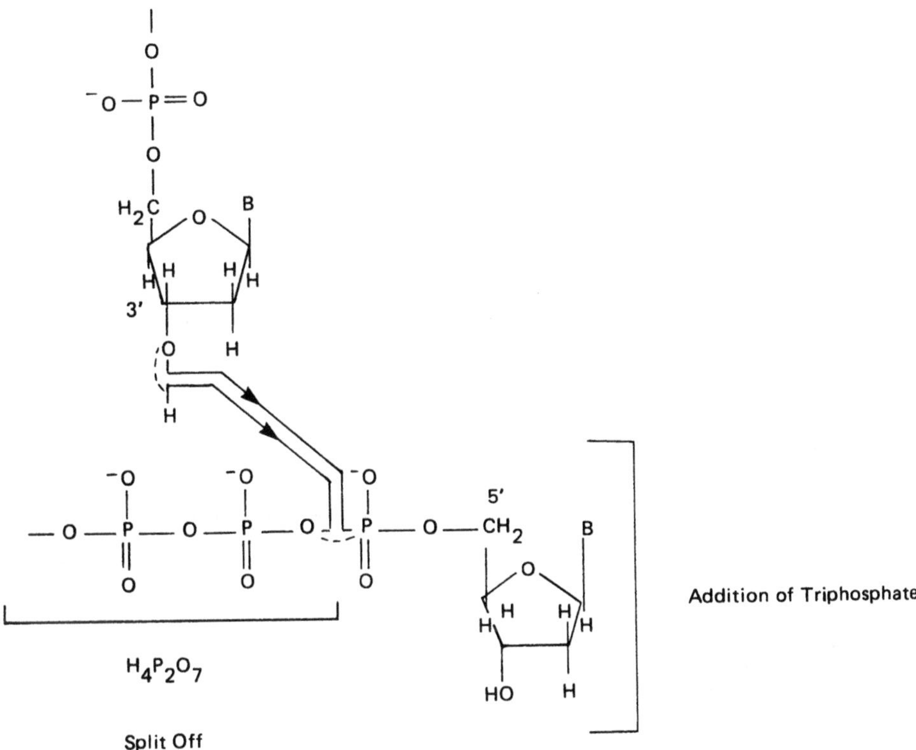

FIGURE 4 Growth of DNA chain. Addition of a triphosphate nucleotide at the 3′ end of the growing DNA chain. Growth is in the 5′ to 3′ direction.

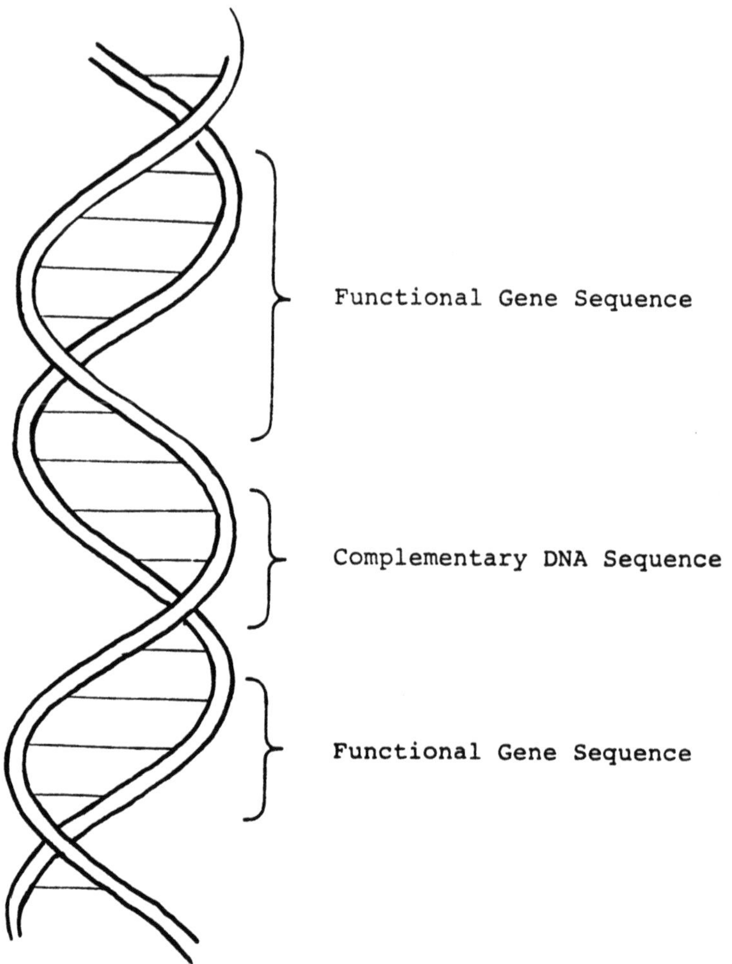

FIGURE 5 Helix formation in DNA. Right-handed DNA helix with alternate sections of functional genes interrupted by a noncoding sequence.

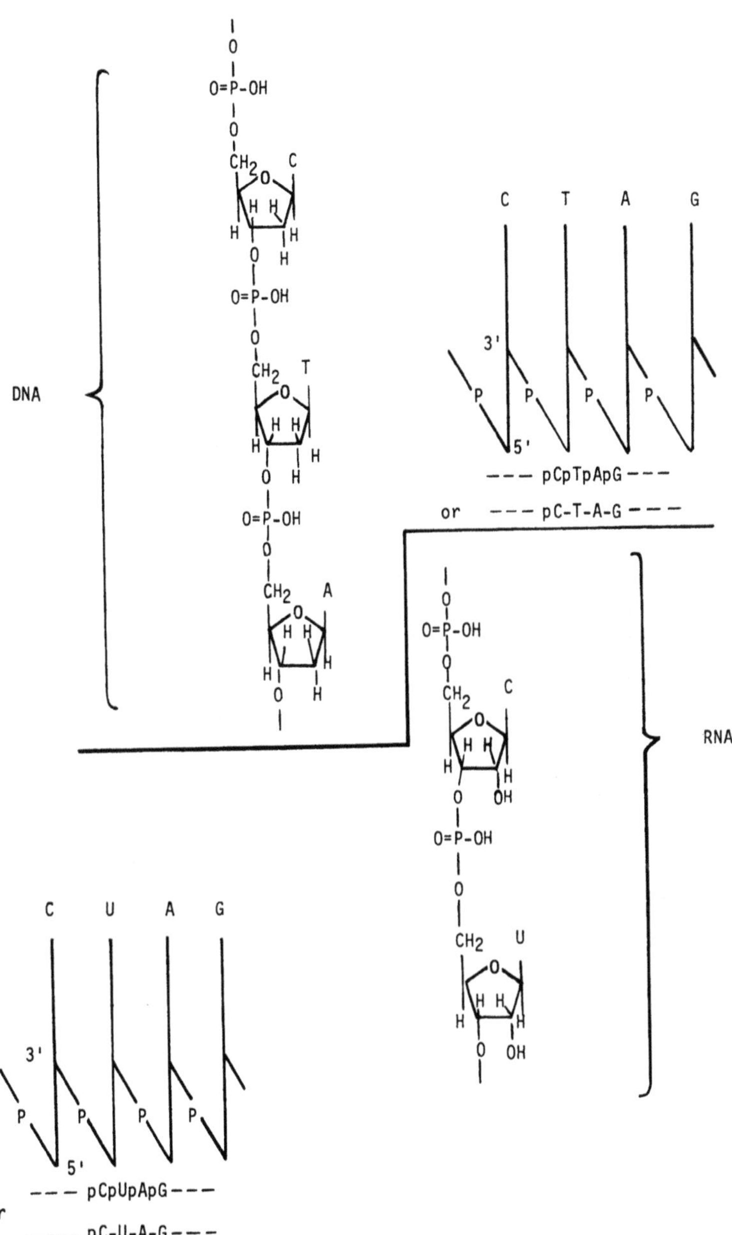

FIGURE 6 Shorthand notation. Convention used in presenting simplified structural formulas by letter symbols. DNA on left side, RNA on right side.

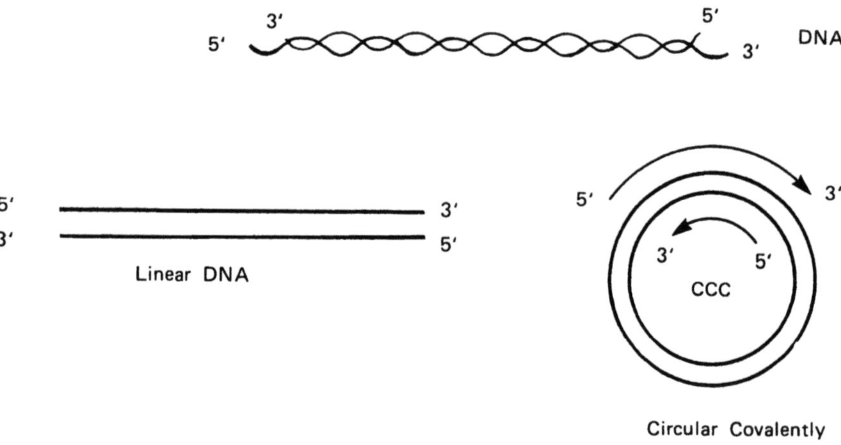

FIGURE 7 Shorthand notation. Simplified graphical presentation for double-stranded DNA.

meaning phosphate at the 3′ location. Also, poly A^+ means multiple A's at the 3′ end.

The simplified graphic presentation of linear and circular double-stranded DNA is illustrated in Figure 7.

REFERENCES

1. Stryer, L. (1981) *Biochemistry*, 2nd ed. San Francisco, CA.
2. Dagani, R. *Chemical & Engineering News. 60*:23–27 (1982).
3. Adams, RLP, et al. (1981). *The Biochemistry of the Nucleic Acids*, 9th Ed. Chapman and Hall, New York.

2
The Cell

To understand the principles of genetic engineering it is necessary first to review the composition and functions of the cell. Living organisms can be divided into two major groups, prokaryotes and eukaryotes. There are several basic differences between these two groups, which are represented, in part, in Table 1 and Figure 1.

Generally, eukaryotes are larger and more complex than prokaryotes, containing complex organelles such as mitochondria and chloroplasts, and also carry multiple chromosomes, as compared with the single chromosome of prokaryotes. Yet despite all these differences, the genetic material, DNA, of both prokaryotes and eukaryotes is identical in its basic structure and function. This enables the researcher to take DNA from one living organism, to put this DNA into an appropriate "vector," and to express the information encoded by this DNA in a completely unrelated organism.

All living organisms comply with the "central dogma," proposed by Crick (1), which states that (a) the genetic information of the cell is carried in its DNA; (b) the DNA serves as a template for the synthesis of a complementary RNA strand (transcription); (c) proteins are synthesized on ribosomes using a specific RNA (messenger RNA) as a code to determine the order of the amino acids that comprise the protein (translation); and (d) the DNA serves as a template for its own reproduction (DNA replication). It was originally thought that transcription and translation were irreversible, until it was found that RNA can serve as a template for the synthesis of DNA by the enzyme reverse transcriptase. This reaction has commonly been used in genetic engineering to convert messenger RNA (mRNA) to DNA, and the product of the reaction has been termed complementary DNA (cDNA).

TABLE 1 Characteristics of Organisms

Feature	Prokaryote	Eukaryote
Structural unit	Bacteria, algae	Unicellular organisms, multicellular plants and animals
Size	<1-2 x 1-4 μm	>5 μm in width or diameter
Location of genetic material	Nucleoid, chromatin body, or nuclear material (different names for same area in which DNA is concentrated)	Nucleus, mitochondria, and chloroplasts
Chromosomes	1	>1
Nuclear membrane	Absent	Present
Histones	Absent	Present
Nucleolus	Absent	Present
Mitochondria	Absent	Present
Chloroplasts	Absent	Present
Golgi apparatus	Absent	Present
Endoplasmic reticulum	Absent	Present
Ribosomes	70S structure	80S structure
Site of electron transport	Cell membrane	Organelles
Genetic exchange mechanisms	Conjugation (plasmid-mediated, unidirectional) Transformation Transduction	Gamete fusion

THE PROKARYOTIC CELL

Most genetic engineering involves manufacturing recombinant DNA and introducing it into appropriate bacteria to produce the desired product. Bacteria are particularly suitable for this purpose because they are small unicellular organisms that can grow in simple, inexpensive media and they rapidly reproduce (generation times of commonly used bacteria such as *Escherichia coli* or *Bacillus subtilis* are 30–60 min with most growth media). A comprehensive two-volume set has

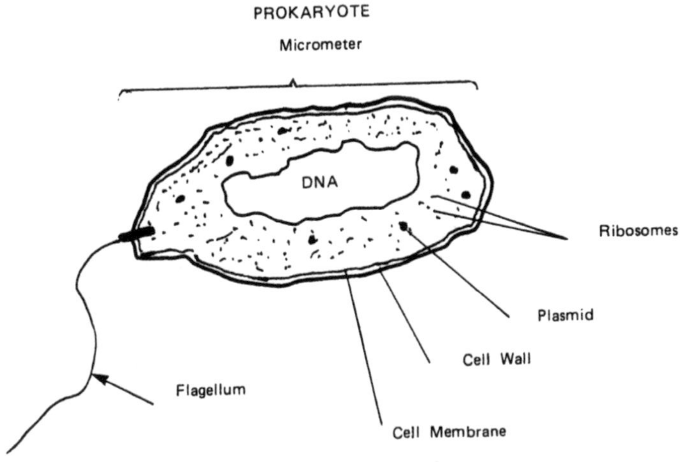

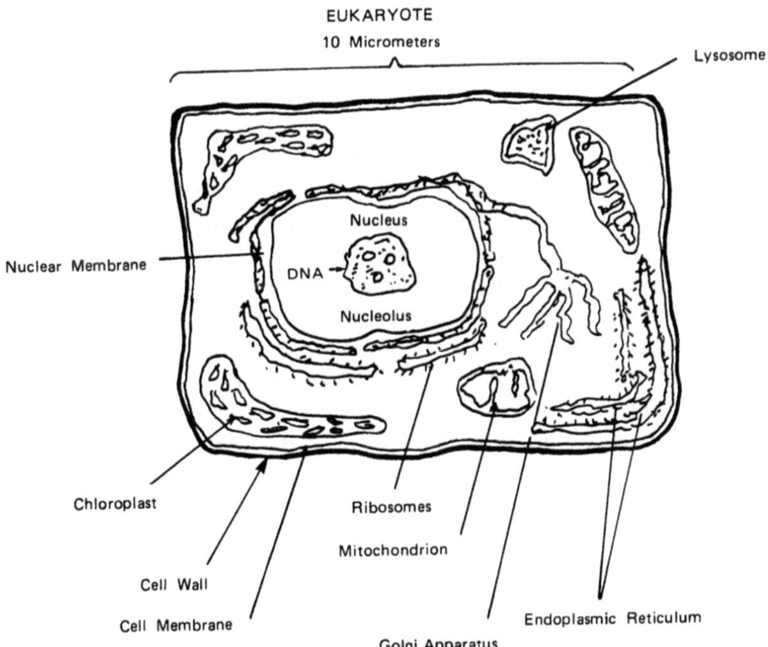

FIGURE 1 Schematic of prokaryote and eukaryote cells. The fundamental structural difference of these two cell types lies in the absence of a nucleus in the prokaryote and in the pronounced difference in cell size.

recently been published (2), comprised of 169 contributions from eminent authors covering all aspects of biochemical, physiological, and morphological diversity of prokaryotic life.

The Cell Wall

Bacteria are divided into two major groups, Gram-positive and Gram-negative, based originally upon the nature of staining of their cell walls (2). Gram-positive cells (Fig. 2a), have a thick cell wall layer, forming approximately 50% of the dry weight of the cell, composed of peptidoglycan and accessory polymers that are usually highly negatively charged (e.g., teichoic acid, a glycerol phosphate, or ribitol phosphate polymer). In Gram-negative cells (Fig. 2b), the peptidoglycan layer comprises only about 5% of the cell mass, and the cells also have an outer membrane layer. The outer membrane is an asymmetric bilayer, the outside of which is composed exclusively or predominantly of lipopolysaccharide while the inside is phospholipid. This bilayer typically contains a few (2–5) major and many (30–50) minor protein species. The functions of the major proteins are to maintain the structural integrity of the outer membrane and/or to form pores in the membrane through which essential nutrients may pass.

Although the cell walls of Gram-positive and Gram-negative bacteria are very different, the peptidoglycan of all bacteria (except *Mycoplasma* and *Archaebacteria*) is quite similar. It is composed of a backbone of alternating *N*-acetylglucosamine and *N*-acetylmuramic acid residues linked by β-1,4 glycosidic bonds. Attached to the lactyl group of *N*-acetylmuramic acid is a tetrapeptide containing alternating L- and D-amino acids. The most commonly occurring tetrapeptide sequence is L-alanine-D-glutamate-L-lysine (or mesodiaminopimelic acid)-D-alanine, with the bond between D-glutamate and the basic L-amino acid involving the γ-carboxyl rather than the α-carboxyl residue of D-glutamate. The ω-amino group of the basic third amino acid is crosslinked, either directly or with a variety of interpeptide bridges, to the carboxyl group of the D-alanine residue of another tetrapeptide. In this manner the peptidoglycan layer is essentially a single macromolecule that forms a net around the bacteria, making the cell osmotically stable and conferring the typical shape of the organism (e.g., rod, coccus).

To obtain either intracellular protein products or recombinant DNA from the bacterium, it is necessary to break open (lyse) the cell, particularly to break the peptidoglycan net. Cells can be broken by mechanical shearing, ultrasonic rupture, or by enzymatic degradation of the peptidoglycan (e.g., with lysozyme). Gram-positive cells can be lysed simply by enzymatically degrading the peptidoglycan in hypertonic media; Gram-negative cells can retain structural integrity even if the peptidoglycan layer as been disrupted, and solubilization of the outer membrane with detergents is often necessary to lyse the cell. Following rupture

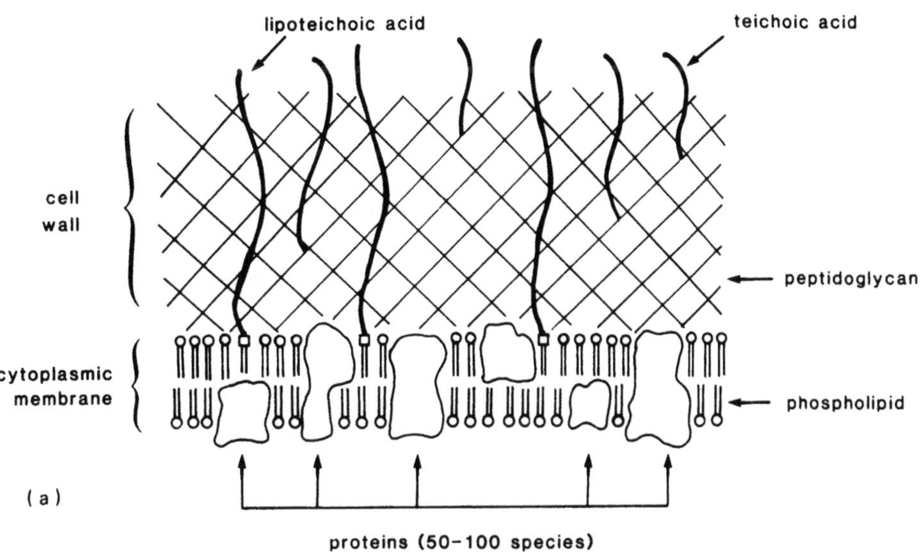

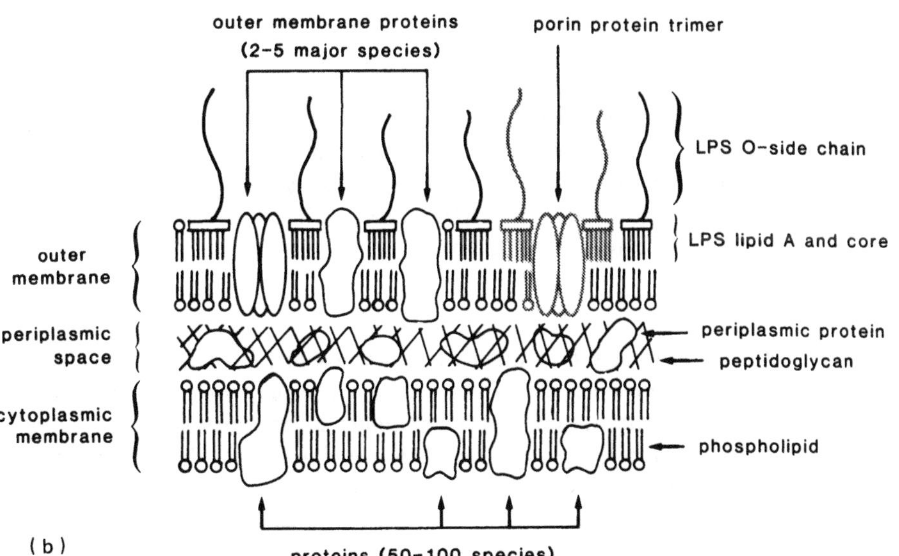

FIGURE 2 Diagrammatic presentation of cell wall structures. (a) Cell wall cross section of Gram-positive bacteria. (b) Structure of Gram-negative cell wall.

of the cells the wall fragments can be separated from the soluble protein and DNA fractions by sedimentation in a centrifuge (usually at approximately 1,000 x g).

The Cell Membrane and Cytoplasm

Bacterial cells contain a cytoplasmic membrane that serves multiple functions including its role as a physical barrier, and as a site for enzymatic reactions such as DNA synthesis, oxidative phosphorylation/electron transport reactions, and vectorial reactions such as active transport of extracellular solutes against a concentration gradient. This membrane is a lipid bilayer containing predominantly polar phospholipids, especially phosphatidyl ethanolamine, phosphatidyl glycerol, and cardiolipin. The membrane contains about 20% of the total cellular protein and has a protein/lipid ratio ranging from 2.0 to 4.0, a higher ratio than is observed in eukaryotes. There are 50 to 100 different proteins found in a typical bacterial cytoplasmic membrane, but none of these are glycoproteins, which are extremely common in eukaryotic membranes.

The cytoplasm of bacterial cells contains about 1000 different protein species, a variety of ribonucleic acids including 3 ribosomal RNA, 30-60 transfer RNA, and 400-500 messenger RNA species, and a single chromosome. The ribosomal RNA species combine with ribosomal proteins (53 different proteins in *E. coli*) to form ribosomes, which are free in the cytoplasm of bacteria and assemble on mRNA to form polysomes. The bacterial chromosome has a closed loop configuration (covalently closed circular) and is negatively supercoiled. This single chromosome contains all of the genetic information essential to the growth of the organism. Although there are no physical compartments within the bacterial cytoplasm, there is a DNA domain, called the nucleoid or nuclear body, which is visible in electron micrographs.

Plasmids

Some (but not all) bacteria contain extrachromosomal DNA that is smaller in size than the chromosome. This DNA, called a plasmid, is circular and supercoiled, and it replicates autonomously. Plasmids do not carry genetic information essential for bacterial growth, but they may carry genes that give the bacterial host a selective advantage. These genes include those coding for antibiotic resistance, toxin production, or ability to use additional organic compounds as a sole carbon source. Conjugal plasmids contain the genetic information to catalyze their transfer from one bacterium to another. If the plasmid carries no known phenotypes, it is *cryptic*. Plasmids are extremely important in recombinant DNA as *vectors* which carry foreign DNA into the bacterium. Figure 3 is a microscopic view of a plasmid.

Bacterial strains can harbor more than one plasmid (coexisting plasmids) as

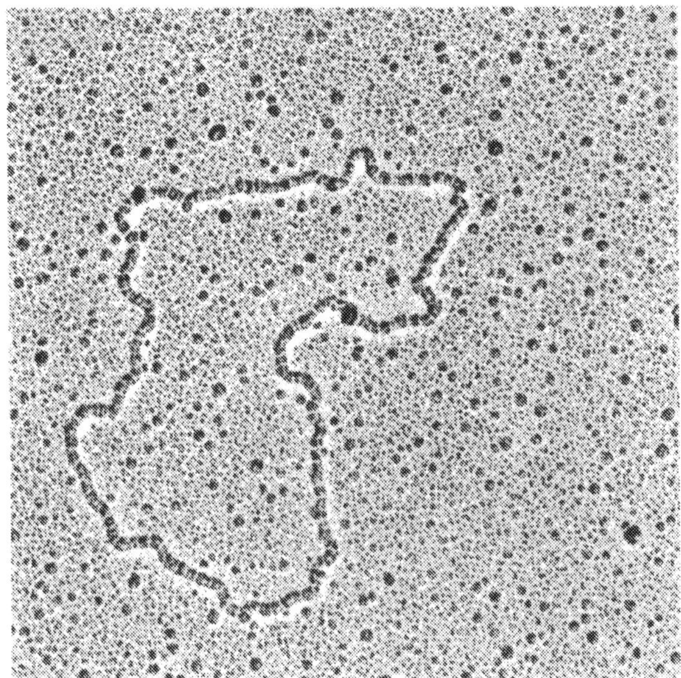

FIGURE 3 Electronphotomicrograph of plasmid. The closed (circular) structure of a plasmid from *E. coli.* Magnification is 100,000 x. (Courtesy J. E. Donelson, Biochem. Dept., University of Iowa).

long as they belong to different incompatibility groups. Two plasmids of the same incompatibility group cannot coexist in the same cell, apparently because of competition for the plasmid replication functions. Over 25 different incompatibility groups have been identified for *E. coli.*

Another plasmid characteristic, called host-range function, is the ability to exist in more than one species of bacteria. Plasmids that can be maintained in a large variety of bacterial species are called broad host-range plasmids. These plasmids can be extremely useful in genetic engineering experiments because they can be used to construct "shuttle vectors," allowing the cloning of DNA in a host such as *E. coli* and its subsequent expression in another bacterial strain (see Chap. 6).

Bacteriophage

Viruses that infect bacteria are called bacteriophage. As with other viruses, bacteriophage are incapable of reproducing autonomously and are dependent

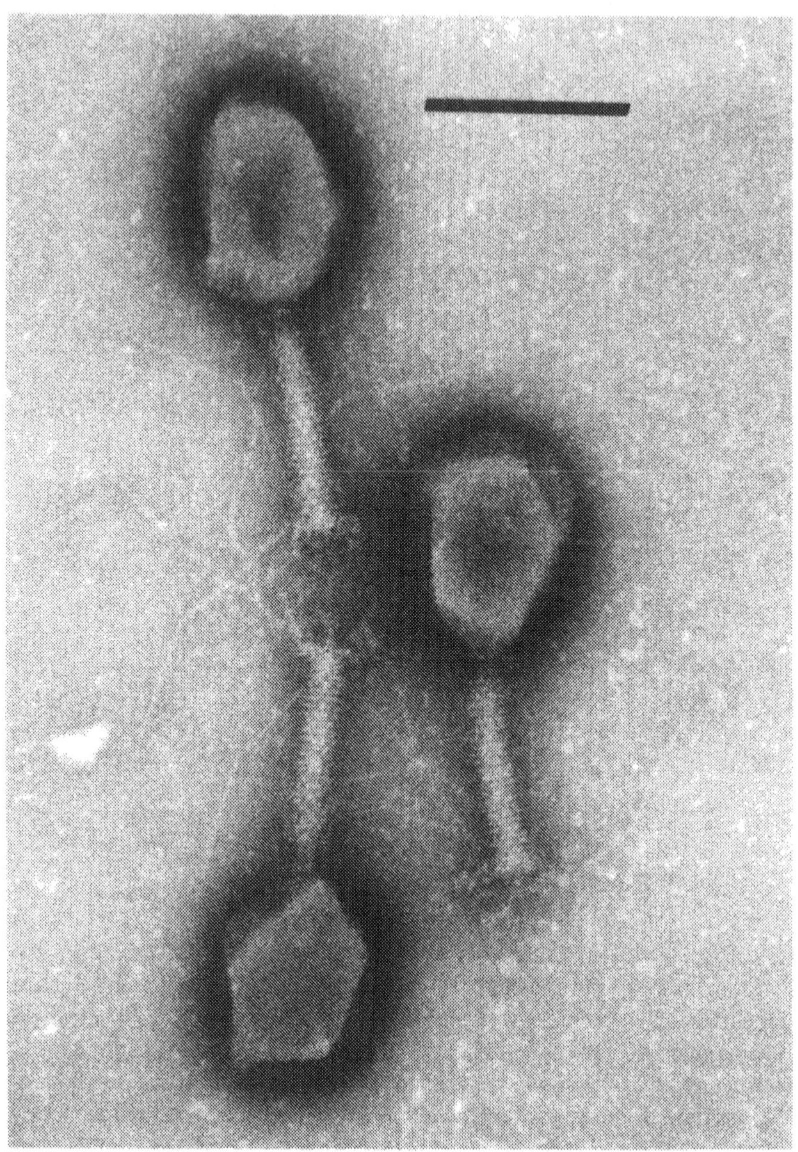

(a)

FIGURE 4 Electron micrograph of Virus. (a) A cluster of bacteriophage T4 virus particles. Negative staining with sodium phosphotungstate. Each virus particle consists of a head capsid containing the DNA and a tail consisting of a contractile sheath to which are attached a base plate and long tail fibers which facilitate attachment to the host cell. The magnification represents 0.1 μm. Fibers appear as barely visible white streaks. (b) Conventional dimensional diagram of a virus particle. Dimensions in Angstroms. (Courtesy of Dr. Donald H. Walker, Professor of Microbiology, University of Iowa).

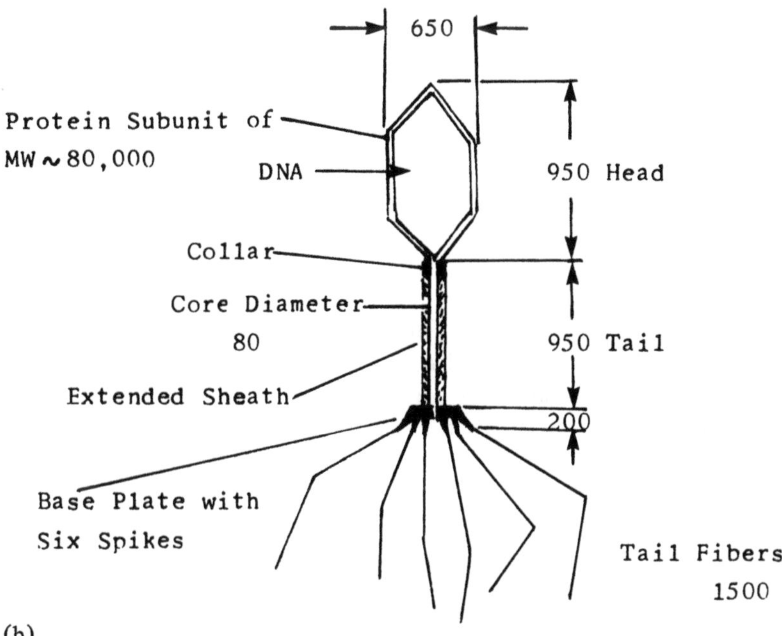

(b)

upon host cell functions for their replication. The genome of bacteriophage can be composed of either DNA or RNA, and the DNA can be either single or double stranded. Bacteriophage can be lytic, meaning that infection causes the death and lysis of the bacterial cell, or they can be lysogenic (or temperate), in which case the virus is replicated in concert with the bacterium and most of the genes of the virus are not expressed (repressed). A lysogenic phage can be induced by a variety of means so that it becomes a lytic phage. Bacteriophage can also be very useful as *vectors* and have the advantage that larger pieces of DNA can be cloned into them than into plasmids. Figure 4 illustrates the appearance of a virus and its structure.

For a bacterium to be infected by a bacteriophage it must contain appropriate receptors on its cell surface to which the phage binds. Loss of these receptors by mutation renders the bacterium resistant to the phage. These bacterial receptors usually have a function which is beneficial to the bacterium; for example the *lam*B receptor for the bacteriophage λ is required for the cell to utilize maltose as a carbon source. This requirement for a specific receptor, which often involves more than one surface macromolecule, naturally limits the number of bacterial species which a given bacteriophage can infect.

In addition to becoming resistant to bacteriophage by altering the phage receptor, a bacterium can protect itself from infection by producing restriction-modification enzymes. These enzymes recognize a specific DNA sequence, usually 4–6 base pairs in length, and cleave the DNA at or near that sequence. The cleaved DNA is then digested by exonucleases. The bacterium protects its own DNA from cleavage by omitting the specific sequence from its genome or by chemically modifying one of the bases in the sequence to render the restriction enzyme ineffective. These enzymes are the basis for the developments in genetic engineering and will be discussed in more detail in Chapter 5.

THE EUKARYOTIC CELL

Eukaryotic cells have a vast variety of shapes and sizes, ranging from the simple unicellular organisms, such as amoeba, to the complex group of differentiated cells found in higher mammals. They are generally larger and more complex than prokaryotic cells. The wide variation in eukaryotic cell morphology cannot be covered here, and this brief overview will be concerned only with the features common to most or all eukaryotic cells.

The most prominent characteristic that distinguishes eukaryotic from prokaryotic cells is the sequestering of cellular DNA in the membrane-enclosed nucleus. The chromosomal DNA appears to be linear rather than circular and is coated with proteins, especially histones. Another feature of eukaryotic cells is the presence of membrane-enclosed organelles, including mitochondria in all aerobic eukaryotes, chloroplasts in photosynthetic eukaryotes, lysosomes, and Golgi apparati. In addition to organelles, there is a great deal of membrane present in the cytoplasm of eukaryotic cells; this membrane is called the endoplasmic reticulum.

Eukaryotic cells also share some similarities with prokaryotic cells. The cell is surrounded by a lipid bilayer containing a variety of proteins; the cytoplasm contains ribosomes; and the genetic information of eukaryotes is contained in its DNA.

Nucleus

The eukaryotic cell has a well-defined nucleus surrounded by a double nuclear membrane containing large nuclear pores (100 nm). The space between the two membranes is called the perinuclear space, and the nuclear membrane appears to arise from an extension of parts of the endoplasmic reticulum. The nucleus appears spheroid in electron micrographs of most eukaryotic cells, but in some cases, such as polymorphonuclear leukocytes, nuclear lobulation occurs that appears to give multiple nuclei when the cell is sectioned in preparation for electron microscopy.

Most of the genetic material is located within the nucleus, in the form of chromosomes. When the cell divides, the nucleus replicates and divides before division of the cytoplasm occurs. Since the nucleus contains the DNA of the cell, it is the site of DNA-directed RNA synthesis (transcription). This RNA is then processed and transported through the nuclear pores to the cytoplasm where translation occurs (see Chap. 3).

A nucleolus is usually present within the nucleus of eukaryotic cells. It is the site of synthesis of ribosomal RNA and the site where ribosomal RNA and ribosomal proteins are assembled into ribosomes. The DNA that codes for ribosomal RNA is contained within the nucleolus and is repeated, with several hundred copies per haploid genome.

Chromosomes

The word *chromosome* was coined by an early investigator, Dr. Waldeyer of Germany, in 1888. Chromosomes appear as "colored bodies" and, when condensed during mitosis, are frequently visible under a light microscope. The portions of the nucleus that are stainable are called chromatin, and consist of proteins complexed with DNA. The term *euchromatin* refers to that portion of chromosomes which disappears during interphase (after cell division) and *heterochromatin* refers to that which persists. Euchromatin appears to be extended DNA that is being transcribed.

The prokaryotic cell has only one circular double-stranded DNA molecule within its single chromosome, and therefore, has only one copy of most genes (this is termed haploid). In most eukaryotic cells, the chromosomes are present in two copies, one derived from each parent. This state is called diploid, and the genes that also occur in two copies are called alleles. Some eukaryotic cells are haploid, whereas other contain many copies of each chromosome and are called polyploid.*

Ribosomes and Mitochondria

The *ribosome* is a granular body found in both the eukaryotic and prokaryotic cell. It is responsible for receiving genetic instructions and translating them into production of proteins. Ribosomes can be attached to membranes such as endoplasmic reticulum or they can be dispersed in the cytoplasm of a cell. Sometimes, in the cytoplasm, they form clusters called "polysomes," which contain several ribosomes engaged in the process of translating the same mRNA. A recent publication by Lake (3) reviews the status of ribosome knowledge. A large body of data suggests that the ribosomes in both prokaryotes and eukaryotes consist of two units: a large subunit and a small subunit.

---

*See Glossary for definition.

There are some differences between prokaryotic and eukaryotic ribosomes, the most basic is that prokaryotic ribosomes are smaller than those of eukaryotes, and, therefore, sediment less rapidly in a sucrose gradient during centrifugation (70S vs. 80S).* The prokaryotic ribosome contains three sizes of ribosomal RNA (23S, 16S, and 5S) and 53 ribosomal proteins, whereas the eukaryotic ribosome contains four sizes of ribosomal RNA (28S, 18S, 5.8S, and 5S) and 70 ribosomal proteins. Both kinds of ribosomes consist of two ribosomal subunits (30S and 50S for prokaryotes, 40S and 60S for eukaryotes) and protein synthesis initiates on the smaller (30S and 40S) subunit. A more detailed discussion of protein synthesis is presented in Chapter 4.

The *mitochondria* are double-membraned organelles in which oxidative phosphorylation occurs. Eukaryotic cells that lack mitochondria must derive their energy by fermentation reactions rather than coupling the formation of ATP (adenosine 5′-triphosphate) to cellular respiration.

There can be as few as 1 to 5 mitochondria in some lower eukaryotic cells, as has been shown for yeast. Other extremely metabolically active cells contain thousands of mitochondria. An average mitochondrial dimension is 0.5 to 1.5 μm, although this varies tremendously with cell type. Mitochondria have their own set of ribosomes which are distinct from cytoplasmic ribosomes and are about 70S in size. Mitochondrial DNA is double stranded, *c*ovalently *c*losed, *c*ircular or CCC DNA, which contains a full set of tRNA (*t*ransfer RNA) and rRNA (*r*ibosomal RNA) genes, as well as about 10 genes that code for specific proteins. In mammalian mitochondria it is about 15,000 base pairs; in yeast, it is about 75,000 base pairs.

It has been proposed that mitochondria arose from bacteria that were engulfed by early eukaryotic cells. This hypothesis is based upon the similarities between the 70S ribosomes of mitochondria and bacteria, the sequence of 5S ribosomal RNA, and the sensitivity of mitochondrial transcription and translation reactions to antibiotics that inhibit those reactions in prokaryotes but not in eukaryotes.

Chloroplasts

Photosynthetic eukaryotes contain membrane-bound organelles, called chloroplasts, that carry out the photosynthetic reactions of the cell. Chloroplasts are quite similar to mitochondria in that they have their own DNA which codes for ribosomal and transfer RNAs and a few proteins, and they contain 70S ribosomes. The chloroplast contains chlorophyll and converts light energy into the chemical energy of adenosine triplosphate (ATP) with the evolution of oxygen.

---

*See Glossary for definition.

TABLE 2 Approximate Chemical Composition of a Rapidly Dividing *Escherichia coli* Cell[a]

Component	Percent of total cell weight	Average MW	Approximate number per cell	Number of different kinds
H_2O	70	18	4×10^{10}	1
Inorganic ions (Na^+, K^+, Mg^{2+}, Ca^{2+}, Fe^{2+}, Cl^-, $PO_4{}^{1+}$, $SO_4{}^{2-}$, etc.)	1	40	2.6×10^8	20
Carbohydrates and precursors	3	150	2×10^8	200
Amino acids and precursors	0.4	120	3×10^7	100
Nucleotides and precursors	0.4	300	1.2×10^7	200
Lipids and precursors	2	750	2.5×10^7	50
Other small molecules (heme, quinones, breakdown products of food molecules, etc.)	0.2	150	1.5×10^7	250
Proteins	15	40,000	10^6	2000–3000
Nucleic acids				
DNA	1	2.5×10^9	4	1
RNA	6			
16S rRNA		500,000	3×10^4	1
23S rRNA		1,000,000	3×10^4	1
tRNA		25,000	4×10^5	60
mRNA		1,000,000	10^4	1000

[a]Weight 10^{12} daltons.
Source: From Ref. 5. Reprinted by permission.

It has been proposed that chloroplasts also arose from bacteria, in this case from photosynthetic bacteria.

Golgi Apparatus

The Golgi apparatus is composed of parallel stacks of flat sacs or vesicles. It is present only in eukaryotic cells and is located near the nucleus. The main function of the Golgi apparatus is to collect and concentrate secretions formed in the cell, package them into storage granules, and release them outside the cell.

Lysosome

A lysosome is a membrane-bound granule containing the enzymes which break down certain cellular material. Lysosomes are the "digestive system" of a cell.

Chemical Composition of Cell

Informationally, the composition of an *E. coli* cell is of interest. A summary of main components is presented in Table 2. As indicated in the table heading, cell composition will vary somewhat as the cell cycle changes; that is, from a newly formed cell to a cell approaching the dividing stage. The terms rRNA, tRNA, and mRNA are discussed in more detail in Chapter 3. Some informative matter is also presented Stent's *Molecular Genetics* (4).

REFERENCES

1. Crick, FHC. On protein synthesis. *Symp. Soc. Exp. Biol.* 12: 138–163 (1958).
2. Starr, MP, Stolp, H, Trüper, HG, Balows, A, Schlegel, HG (Eds), (1981). *The Prokaryotes: A Handbook on Habitats, Isolation, and Identification of Bacteria* Springer Verlag, Berlin-Heidelberg, Vols. I and II.
3. Lake, JA. The ribosome *Sci. Am.* 245 (2): 84-97 (1981).
4. Stent, GS. (1971). *Molecular Genetics*. W. H. Freeman & Co., San Francisco. (1976).
5. Watson, JD. (1976). *Molecular Biology of the Gene*, 3rd Ed. W. A. Benjamin, Inc., Menlo Park, CA.

3
DNA, RNA, and Genes

The characteristics of all organisms are determined by genes. These are represented by specific sequences of the "nucleotide bases" on the polymer strands of DNA and RNA, that is: deoxyribonucleic acid and ribonucleic acid. The acid character of these polymers arises from the ionization of the OH groups located on the phosphodiester bridges of the structure.

Watson and Crick (1) in 1953 described a most imaginative interpretation of the X-ray crystallographic patterns of DNA fibers. They inferred that it consisted of double strands twisting around an invisible axis (a spindle) as shown in Figure 1. A later personal account was presented by Watson (2). While DNA within the organism always exists in a double-stranded helical formation, RNA almost always occurs as a single strand. The normal coiling configuration of DNA is such that a complete turn of the double helix contains about 10 base pairs. One strand is considered to be the template strand, while the second strand is the complementary strand.

A good introductory textbook is *The Biochemistry of The Nucleic Acids* by Adams et al. (3), which considerably details the composition, alternate structures, and allied compounds, as well as pertinent reaction behavior. However, a fairly good understanding of biochemical phenomena is needed to thoroughly appreciate this text. Variations in nucleic acid structrues are described from a stereochemical viewpoint by Neidler (4). Also of interest should be *The DNA Story,* a recent volume by Watson and Tooze (5). Its main thrust is, however, concerned with the ethical, legal, and potentially political aspects of recombinant DNA developments. Thus, the major part of the presentation is devoted to documentation of events, and pertinent correspondence and legal papers are reproduced in full.

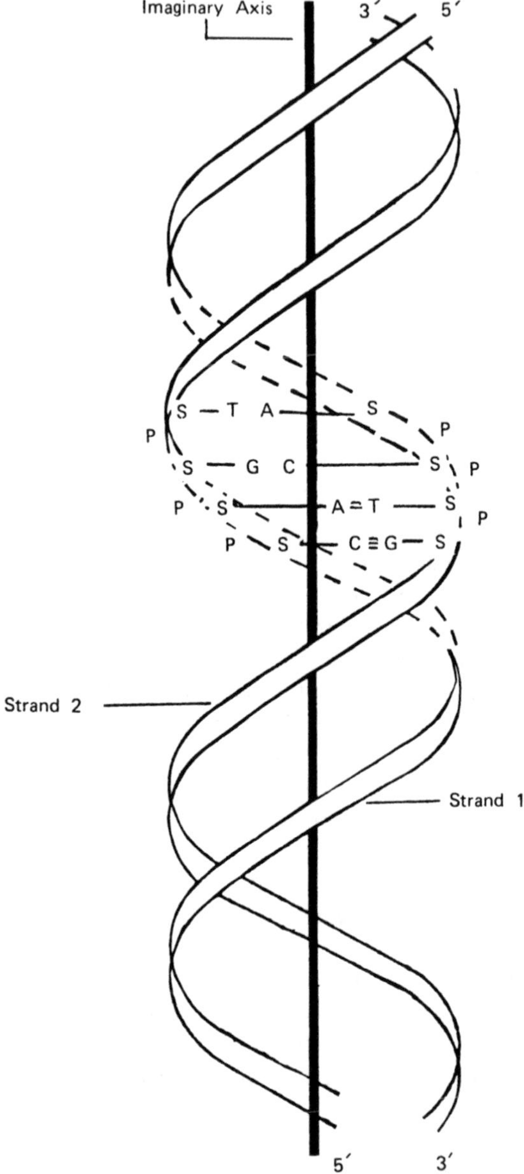

FIGURE 1 Diagram of DNA helix. Double-stranded DNA is visualized to exist in a spiral configuration, mostly in a right-handed direction. Pairing of base pairs is A-T and C-G. P = phosphorus; S = sugar (deoxyribose); A, C, G, and T= bases.

HISTORY OF DNA

During a discussion with graduate students in the biochemistry laboratory, I stated that DNA was not even known to exist when I attended biology lectures in the early 1920s. Whereupon an astute student remarked that it was indeed known. He left the group and returned shortly with literature documenting the early beginnings of DNA research. A rather concise historical sketch is presented by Adams et al. in their introduction (3).

Chronologically, the steps in understanding the development in DNA, and one should add, in RNA, are summarized as follows:

1868 Friedrich Miescher (with Hoppe-Seyler in Tübingen) isolated pus cells from discarded surgical bandages and found that the cells contained an unusual phosphorus-containing compound, which he named "nuclei" (today's nucleoprotein). This constitutes the basic discovery of the existence of an as-yet vague constituent of cells.

1872 Miescher continued his studies at the University of Basel, and in 1872 he reported that sperm heads, isolated from salmon sperm, contained an acidic compound (today's nucleic acid) and a base which he named "protamine." This represents the beginnings of the unravelling of a vital cellular component.

1889 Altman continued Miescher's investigation. It had been recognized that nucleic acids were normal constitutents of all cells and tissues. Altman described a method for preparing protein-free nucleic acids from animal tissue and yeast. By now numerous biochemical investigations had become of interest and continual parallel studies flourished in Europe and the United States.

Note that this step constituted the important recognition that nucleic acids in the organisms are frequently enmeshed with proteins.

As interest in these fascinating compounds grew, research activities accelerated and expanded. The number of investigators multiplied rapidly; so did the number of significant findings. A very useful source of nucleic acid was discovered to be the *thymus gland*, and a great deal of effort was expanded on related analyses. For instance, hydrolysis of the medium showed the presence of the *purines*, adenine and guanine, as well as the *pyrimidines*, cytosine and thymine, and the presence of a deoxypentose (deoxyribose) and phosphoric acid.

1920 Here is an example which typifies the intricate process of research. Findings must be interpreted on the basis of existing knowledge, and extended through rational speculations. Thus, the state of the art indicated the existence of of two DNAs, one represented in the mammalian, or perhaps the whole animal kingdom, and the other as represented in plants.

Such conclusions seem entirely appropriate in light of the then existing knowledge. Adding to this the difficulties encountered by the early investigators who had to cope with inadequate and laborious analytical methods and the problems posed by extraction and purification of the nucleotides, one must be awed and impressed by their accomplishments. This situation and the dualistic belief continued for some 30 years, and a didactic statement by Jones in 1920 seemed to firm up this position.

1940. Inevitably some investigators found that the pentose derivatives were present in animal tissues and so was uracil—conversely, deoxyribose and thymine were identified in plant tissues. It was not until the early 1940s that the existance of both DNA and RNA in animal and plant cells alike came to be taken for granted.

1950. From then on the discoveries and elucidation of structures accelerated, climaxing with Watson and Creek's proposal of the DNA double-helical conformation in 1953.

1970. New modern and sophisticated analytical and preparative methods permitted complete identification of DNA and RNA and definition of their interrelationship, as well as their respective functions. Likewise, the recognition of the ability of specific enzymes to cut polynucleotide strands at specific locations marked a milestone in the discovery of recombinant techniques. Most recently, the preparation of synthetic nucleotide sequences led to the development of "gene machines," which now make it possible to synthesize any number of derived sequences.

GENES

The units of heredity in living organisms are called genes. The biochemical definition of a gene is that it is a DNA sequence that contains the information that leads to the synthesis of a single protein or a single RNA product. In prokaryotes, genes may be organized into single regulatory units called operons. An operon typically contains more than one gene, a promoter where messenger RNA (mRNA) synthesis begins, a terminator where mRNA synthesis ends, and an operator where a regulatory protein determines the rate of initiation of mRNA synthesis. The mRNA that is synthesized from the operon contains the sequences of all of the genes of the operon and is called polycistronic.

The site at which RNA polymerase binds to DNA and initiates the synthesis of messenger RNA is called a promoter. There are four well-defined promoters used in recombinant procedures. These are the tryptophan (*trp*) promoter and the *lac* promoter of *Escherichia coli* (the latter usually representing the *lac*Z gene, β-galactosidase), the promoters for the phage lambda *N*-gene, and the ampicillin resistance gene (Amp^r), which codes for the β-lactamase enzyme in the plasmid pBR322. The abbreviated nomenclature for these promoters is: P_{trp}, P_{lac}, P_L, and $P_{\beta\text{-lact}}$, (Amp^r).

The *lac* promoter is probably the one used most frequently, and so a few pertinent comments are in order. A special region of DNA adjacent to the promoter region of the gene exercises control over initiation of gene transcription. This region is called the *operator*, as it determines the rate at which the three genes of the *lac* operon will be transcribed. One of these, the *lacZ* gene, codes for the enzyme β-galactosidase, which catalyzes the breakdown of lactose to glucose and galactose. The combined structure of the operator and the three associated genes is the *operon*. The processes regulating the expression of operons are complicated by the interaction of promoters and operators so that "and/or" situations can arise. In this connection it would be well to consult texts such as Watson's (6) or Keeton's (7). The fairly extensive treatment to thoroughly explain the phenomena is beyond the scope of the present text.

The sequence of bases in specific regions of the DNA comprise the genes of an organism. Such regions are called coding sequences. In eukaryotes, other sections, which seem to have no genetic function and are frequently called "nonsense DNA," "spacer DNA," or "intercistronic DNA," sometimes alternate with the genes (Chap. 1, Fig. 4). The complete complement of DNA in a cell carrying all of that cell's genes is called a genome. The phenomenon of noncoding sequences inserted into some eukaryotic genes and called intervening DNA or introns, has recently received much attention (8,9). When the introns are excised from the messenger RNA, the resulting coding sequence is called an exon.

DNA

As a chemical substance, DNA can be precipitated from its solution, filtered, dried, and stored in a reagent bottle. While it has no life itself, without it, life as we know it, would not exist.

The conventional assumption has been that DNA has a right-handed spiral configuration, (there are *A*, *B*, and *C* forms). There is now reliable evidence that left-handed DNA structures exist. Rich and co-workers in 1979 reported the evidence for a left-handed DNA spiral, named the *Z-DNA* form (10). Rich suggests that Z-DNA may be one of the elements that regulate gene transcription. The current status of Z-DNA was reviewed by Norman (11). It is now thought that both right-handed and left-handed helices can occur in the same two strands of DNA.

DNA Configuration

DNA and RNA chains can undergo a variety of structural configurations of *spatial* character. Actually, one must envision that these polymer chains are capable of dynamic motion. Such motion can be influenced by physical and chemical factors, such as temperature, pH value, enzymes, ionic salt concentration, among others.

In aqueous solution DNA and RNA are soluble at normal concentrations over a wide pH range. Both polymers can be precipitated in a sodium chloride-alcohol medium; a conventional mixture is: 1 volume DNA solution, 0.1 vol 3 *M* NaCl and 2.5 vol of cold ethyl alcohol. If rather dilute DNA solutions are treated, cooling to -20°C is necessary. The DNA concentration in a dilute solution is in the range of 10 μg/ml. A fairly concentrated solution would contain up to 2 mg/per ml.

An important additional configuration of double-stranded DNA is that it undergoes "super coiling." Several enzymes have been implicated in this process. The activities of one such enzyme, DNA gyrase have been reviewed by Fisher (12). The reaction catalyzed by this enzyme is illustrated in Figure 2. Coiling and supercoiling of DNA were vividly described by J. C. Wang (13) in a discussion of "topisomerases," enzymes which conduct the shifting of topological forms.

Nucleosomes

The elementary subunit of a chromosome structure is called a nucleosome. It constitutes a DNA superhelix wound on a core consisting of histone proteins, so that about 160 base pairs are contained in a complete turn of the helix.

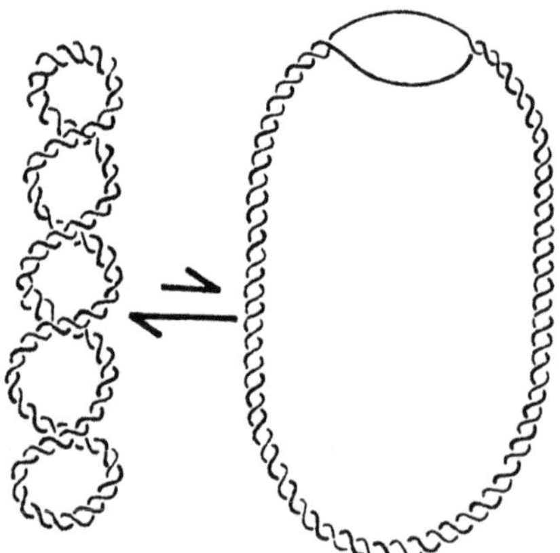

FIGURE 2 Supercoiled DNA. So-called underwound DNA is pictured at right (also relaxed state or negative supercoils) and overwound DNA (positive supercoils) at left. State of supercoiling is determined by enzyme action. Reprinted by permission of Nature, 294 (5842), 607 (1981) McMillan Journals, Ltd., (12).

The substance which forms the chromosomes of higher cells is chromatin. Chromatin itself is an association of the DNA with a substantial number of proteins. The protein portion of the chromatin consists of five types of histone proteins and a large variety of nonhistone proteins. The histone proteins are rich in lysine and arginine, with some histidine. Lysine, for instance, is one of the diaminomonocarboxylic acids and the second amino group confers a strong positive charge on the molecule. It is believed that this basicity interacts ionically with the negative charge of the acid DNA to form a fairly strong bond between the histones and the DNA helix.

The most recent concepts about histone: DNA interactions described by Kornberg and Klug (14), lead to the conclusion that four of the five major histones, called H2A, H2B, H3, and H4, are compacted as wedge-shaped units into a cylindrical structure of about 110 Å diameter and 55 Å height, around which the DNA coils in a tight spiral (Fig. 3). The fifth histone H1 is located at both ends of a histone sequence, so it plays a separate role in sealing off the histone complex. Such substructures seem to represent repeating units (solenoids).

DNA Replication

The mechanics of DNA replication are extensively reviewed in *DNA Replication* by Kornberg (15) There have been many investigations into the mechanical and topological problem of the replication of a double helix, but the most plausible procedure is that of "active unwinding."

Active Unwinding

This concept is shown diagrammatically in Figure 4. The overall process of replication in bacteria involves at least 20 different proteins, where each protein exercises a specific function. Note, that the process requires a rotation of the helix during strand separation, that is, the unwinding action. Some interesting estimates have been made for the rate at which replication proceeds. In *E. coli*, the maximum velocity of the replicating region seems to be about 30 μ/min. This means that 10^5 bases per minute are added to the growing chain. So, an unwinding rate of 10,000 rpm would be required. In mammalian DNA the rates are much lower. Such behavior is another indication of the dynamic nature of living systems. There are no static situations. Another example of continual change is the contention that there is an ongoing exchange of hydrogen atoms in the hydrogen-bonding sites of the paired bases, that is, H atoms on bases exchange with H atoms of the H_2O in the cytoplasm.

The DNA replication of *E. coli* is initiated by a protein labeled *dna*B, and the elongation is controlled by the "DNA-dependent polymerase III."

This enzyme attaches itself to a replication site on the double-stranded DNA which is called an "origin of replication." The replication process involves the

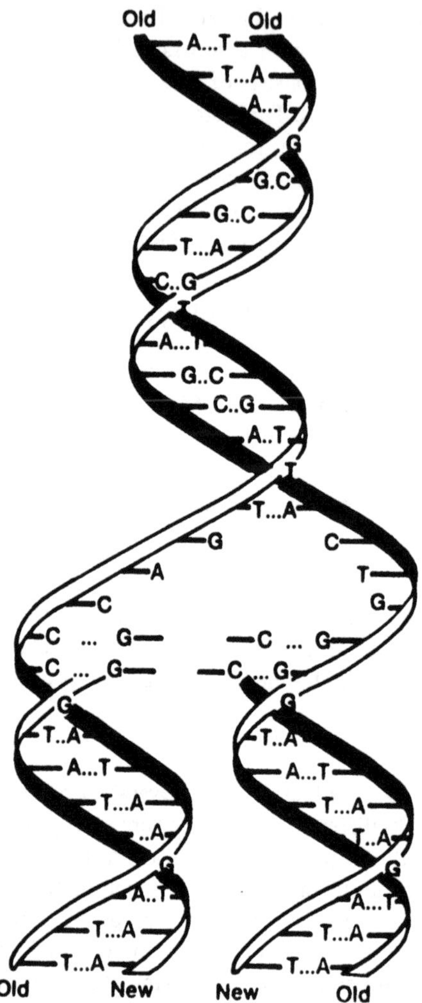

FIGURE 3 Biosynthesis of DNA. Unwinding of ds-DNA to form daughter strands. The double stranded, helical structure of DNA (top), and the uncoiled parent strands and growing, complementary daughter strands formed in DNA replication. Note that each replica consists of one new and one old strand. Reprinted with permission of *Chemical Technology* 16(9): 544 (1986).

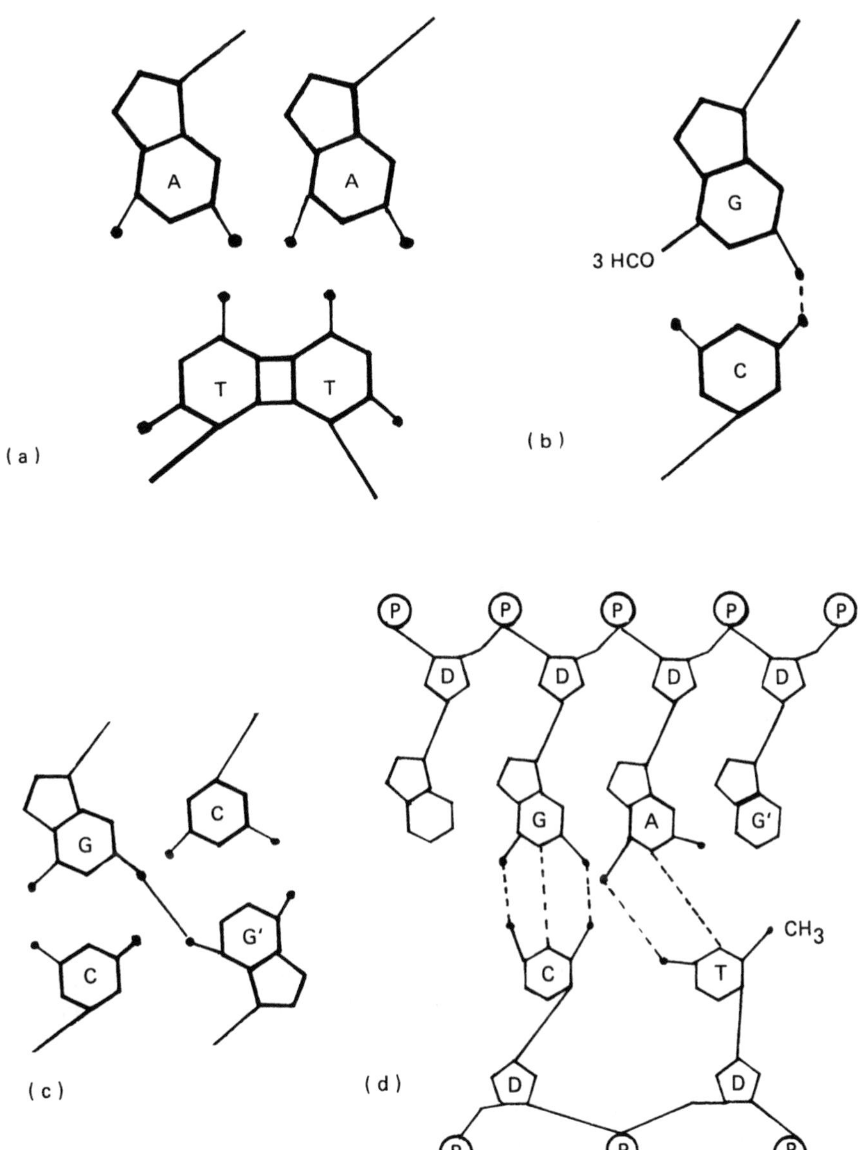

FIGURE 4a-d DNA Faulting. Mutation. Several common cases of breakage or dislocation between DNA strands. See text for detailed explanation.

unwinding of the two strands and addition of short complementary sections whose sequence is directed by the original template strand. The short complementary sections are formed from free nucleotides. As sections are added, the new strand comes into existence to complete the double-stranded polymers (16).

DNA Faulting: Mutation

DNA faulting, leading to respective disturbances in coding, can occur through a multitude of "combining errors." Three such offbeat occurrences are diagrammed in Figure 4. The cases are as follows (17).

Case a: Two neighboring bases, specifically T-T, become bonded internally and thus break two hydrogen bonds with the respective A-A bases. This can occur, particularly with pyrimidines, as a result of ultraviolet and X-ray irradiation.

Case b: A base, for instance a G, can become methylated, resulting in the breakage of its hydrogen bond with the respective C. Methylation of the N-7 position of G or at the O atom in C-6 position is readily accomplished with methyl iodide, dimethyl sulfate, methyl nitrourea, etc.

Case c: Cross-linking of the DNA strands can occur, as for instance G to G. The result is prevention of strand separation leading to cell death. The reaction is accomplished through use of bifunctional alkylating agents such as 1,2-dibromomethane.

Case d: The diagram here is an overall scheme indicating breakage between A and T. The situation shown occurs immediately adjacent to case *c*. The symbol G' is used to relate cases *c*. and *d*.
P = phosphorus
D = deoxyribose

The ultimate effect of such reactions is interference with DNA replication and changes in coding sequences. Unless a DNA repair mechanism can eliminate the "unnatural" changes, they will ultimately result in cell death.

Thermodynamics of Living Matter

The manifold processes which take place in biological systems can be subjected to thermodynamic analysis. Thus, Benzinger (18), published a challenging treatise wherein he described the interpretation of enzyme folding and unfolding, unwinding of the double helix, bond formation, and bond breakage in terms of basic thermodynamic functions; that is, chemical bond energies and heat capacities. Benzinger concludes that the three laws of thermodynamics are as valid as for chemical reactions in general. A complementary text to the physical interpretation of events would be Freifelder's (19). Also Hawker, and Linton's text (20) is a helpful reference.

RNA

The nucleotide units in the RNA molecule are determined by the template strand of the DNA. As previously stated, RNA differs slightly in chemical composition from DNA. Briefly, the thymine base in DNA is replaced by a a uracil base in RNA, and the RNA polymer contains ribose sugar residues instead of the deoxyribose residues of DNA. Also, RNA does not occur in a circular (closed loop) configuration.

The initiation of the RNA chain takes place at a specific site, the promoter region of the DNA. Termination of RNA synthesis similarly is dictated by a terminator region on the DNA. The *transcription* process, DNA to RNA, is controlled by a DNA-dependent RNA polymerase. Such enzymes can be isolated from both prokaryotic and eukaryotic sources. The respective RNA sequences are synthesized in accordance with the base pairing rules, with the above cited restrictions (see Chap. 4, Fig. 1 for diagram).

The overall scheme of RNA biosynthesis is illustrated in Figure 5. The procedure shown holds for prokaryotes and eukaryotes. The role of a specific nucleotide sequence in RNA termination in prokaryotes is presented in Figure 6. Along each gene only the template strand acts as the RNA directive! Synthesis proceeds from the 5′ end of the 3′ end, just as with DNA synthesis. Termination involves the formation of a hairpin loop in the nascent RNA strand and the specific interaction of this loop configuration with the protein *rho*.

Messenger RNA (mRNA)

RNA that is formed from a gene coding for the synthesis of a protein is called messenger RNA. In most or all prokaryotic genes the mRNA that is made can be directly used for translation (see Chap. 4). In eukaryotes, however, many mRNAs are made in precursor form, and intervening, nontranslated portions must be rmoved before translation into the proper gene product can occur. This removal of intron sequences is called processing (see Chap. 4 for details).

Besides processing, two additional structural modifications occur in eukaryotic mRNA. The first, called capping, is the addition of specific nucleotides to the 5′ terminus of the mRNA. The purpose of this structure is not clearly established, and it is speculated that the specific addition to the 5′ end of an RNA molecule protects the mRNA against the action of phosphatases and nucleases, in general. While the normal phosphodiester linkage between the nucleotides in RNA (and DNA) is 5′ to 3′, the cap structure is a 5′ to 5′ linkage through a triphosphate bridge. The structure is illustrated in Figure 7., where the

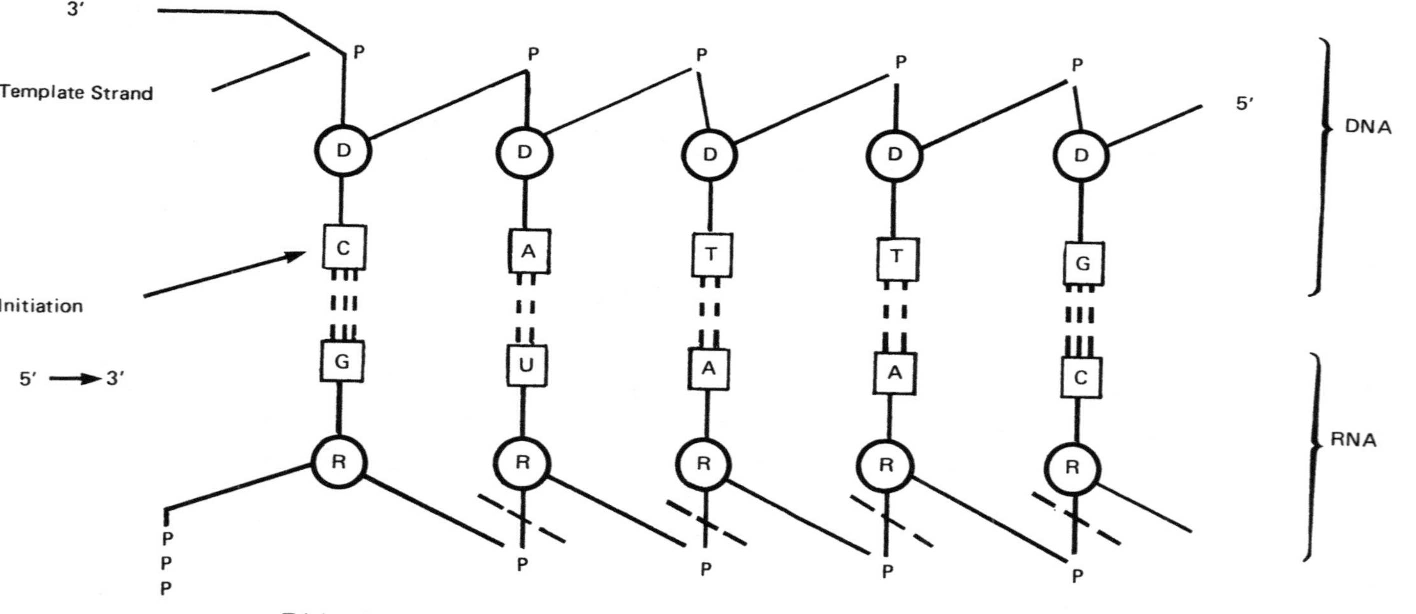

FIGURE 5 Chemical structure of RNA formation. ds-DNA unwinds and template strand codes for RNA; chain growth from 5′ to 3′, by addition of triphosphate complexes. Dashed lines indicate sites of hydrolysis with alkali. P = phosphorus, D = dioxyribose, R = ribose; A, C, G, T, and U = bases.

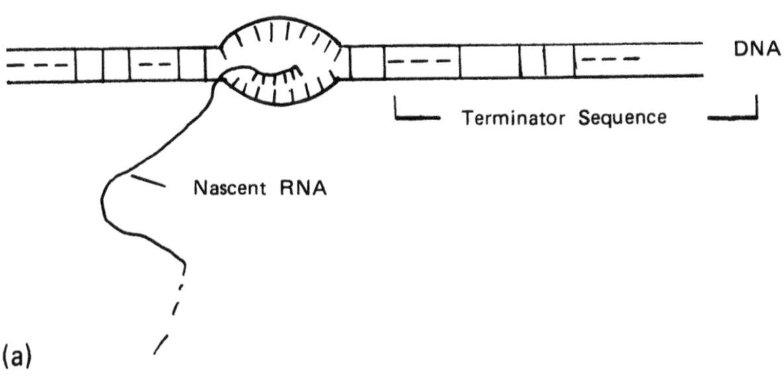

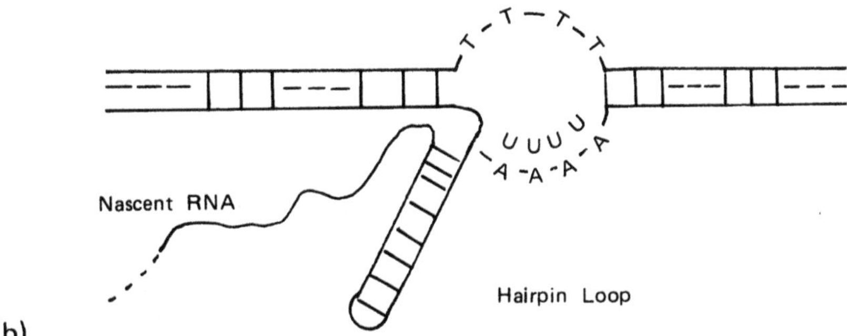

FIGURE 6 Terminator process in RNA formation. (a) Illustrates the terminator sequence in RNA formation. (b) Shows formation of a ternary terminator complex. Note occurrence of hairpin loop structure in terminated RNA.

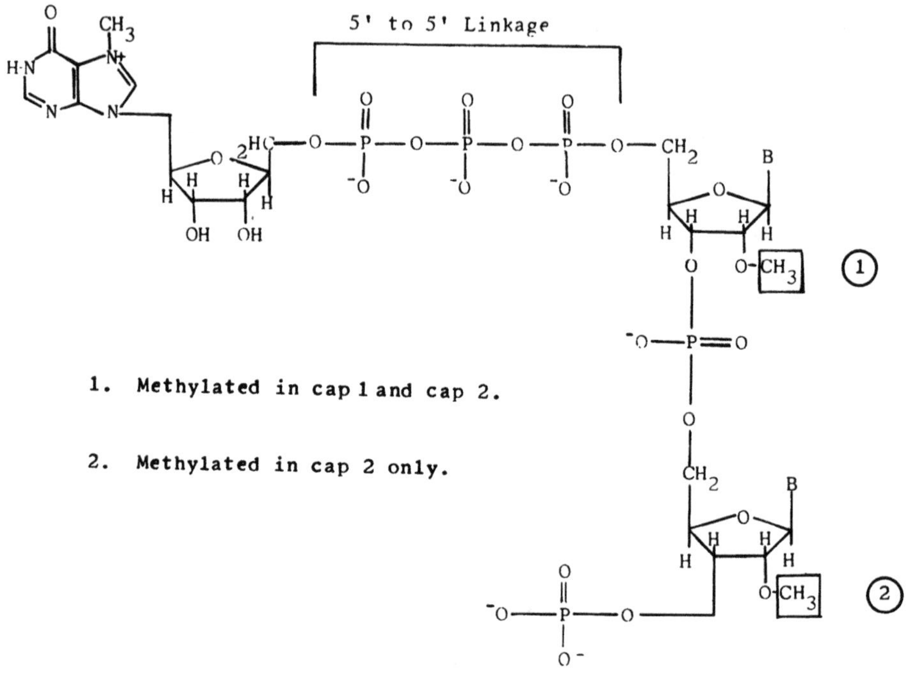

5' to 5' Linkage

1. Methylated in cap 1 and cap 2.

2. Methylated in cap 2 only.

FIGURE 7 Capping of eukaryotic mRNA. Capping to protect ends against enzyme attacks, novel 5′ to 5′ linkage through a triphosphate bridge.

triester linkage is identified. The final base in the cap is a methyl guanine, sometimes labeled *cap* 1 and *cap* 2. *Cap* 1 is always present, while *cap* 2 may be absent. The second modification is the addition of 100–200 AMP residues to the 3′ terminus of the eukaryotic mRNA (polyadenylation). This modification is also thought to protect the mRNA from the action of exonucleases.

Transfer RNA (tRNA)

tRNA is a single-stranded molecule. The tRNA molecules are transcribed from their corresponding genes in the DNA molecule by RNA polymerase III. Activated nucleotides (ribonucleoside triphosphates) must be provided for the synthesis of all RNA molecules including tRNAs.

The basic function of tRNA is to interact with the amino acid and the codon on the mRNA to position the amino acid into its correct location in the polypeptide chain on the surface of the ribosome. tRNA has an anticodon which recognizes the RNA codon. The three bases in the anticodon are complementary in hydrogen bonding to the three bases of the RNA codon (see chapters on Protein Synthesis).

tRNAs form a unique secondary structure termed "cloverleaf," which is illustrated in Figure 8, which shows the structure of yeast tRNA which accepts alanine, $tRNA^{ala}$. There are four base-paired regions and three loops of unpaired regions. Some unusual modified nucleotides are present: they are identified in the figure. These nucleotides are posttranscriptional modifications of the four conventional nucleotides. They are located in unpaired loop regions. The anticodon triplet of importance in protein synthesis is shown as a shaded area in the middle loop. Also, an unpaired tail region of three nucleotides exists at the 3′ end, where the amino acid alanine becomes attached. There are different tRNA species which can accept each of the 20 different amino acids involved in protein synthesis. Some amino acids are bound by more than one tRNA species; multiple tRNAs for the same amino acid are called isoacceptor tRNA. Common characteristics of all the cloverleaf structures are the residue pG at the 5′ end, the anticodon in the middle loop, and the occurrence of the sequence TψCGA in one of the side loops. The full significance of these unique structural characteristics has, as yet, not been elucidated.

Ribosomal RNA (rRNA)

rRNA makes up about 40% of the total mass of ribosome. It is mostly single-stranded but sometimes it doubles back upon itself, making complementary base pairs which do not form along the entire strand. The rRNA molecules act as scaffolds upon which the ribosomal proteins assemble to form ribosomes. At the 3′ end of prokaryotic 16S* rRNA, the Shine-Dalgarno sequence complements the sequences in the mRNA. Hybridization between these sequences binds the mRNA to the ribosome and positions the AUG start codon properly on the ribosome so that the initiator tRNA (fMet-$tRNA_f^{Met}$) can bind to the small ribosomal subunit.

Occurrence of Minor Bases

So-called minor bases do occur in some nucleic acids, but usually in rather small amounts and rarely represent more than 10% of the total base content. The presence of such minor bases, all of which are analogs of A, C, G, T, and U, is more prevalent in RNA structures. In particular, tRNA contains a wide variety of "methylated bases." Table 1 presents a list of the more frequently encountered structures.

---

*See Glossary for definition of *S*, the Svedberg unit.

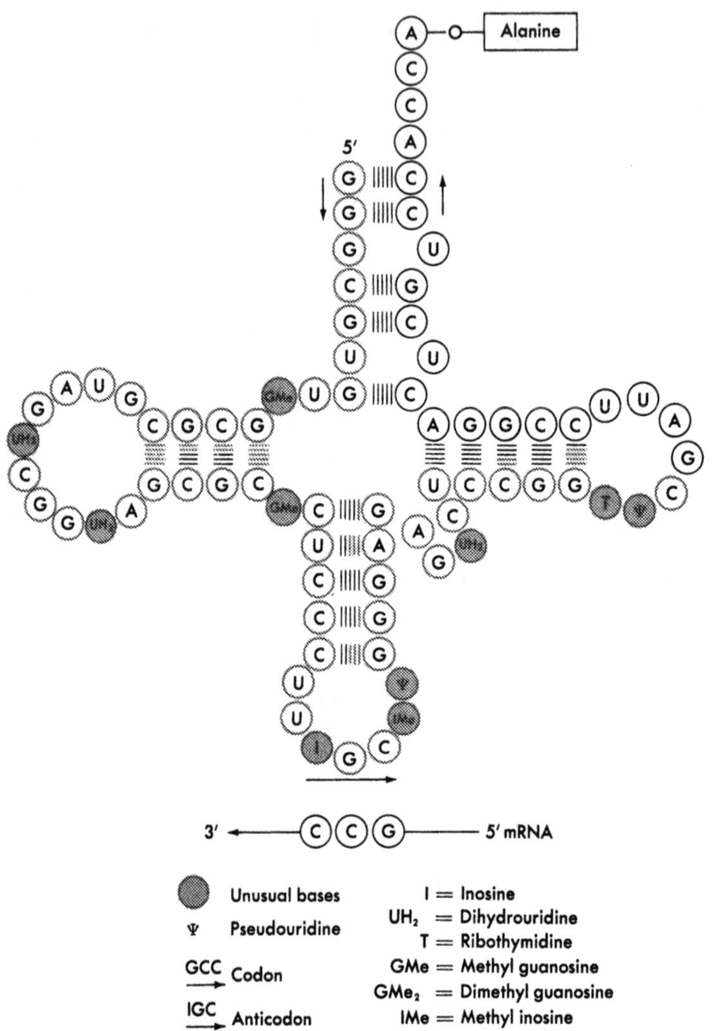

FIGURE 8 Cloverleaf structure of tRNA. The structure shown is for alanine tRNA. Reprinted by permission from J. D. Watson *Molecular Biology of the Gene*, (c) 1976 by the Benjamin/Cummings Publishing Co., Menlo Park, CA (6).

TABLE 1 Some of the More Important Minor Bases in RNA

1-Methyladenine	Dihydrouracil
2-Methyladenine	5-Hydroxyuracil
6-Methyladenine	5-Carboxymethyluracil
6,6-Dimethyladenine	5-Methyluracil (thymine)
6-Isopentenyladenine	5-Hydroxymethyluracil
2-Methylthio-6-isopentenyladenine	2-Thiouracil
6-Hydroxymethylbutenyladenine	3-Methyluracil
6-Hydroxymethylbutenyl-2-methylthioadenine	5-Methylamino-2-thiouracil
1-Methylguanine	5-Methyl-2-thiouracil
2-Methylguanine	5-Uracil-5-hydroxyacetic acid
2,2-Dimethylguanine	3-Methylcytosine
7-Methylguanine	4-Methylcytosine
2,2,7-Trimethylguanine	5-Methylcytosine
Hypoxanthine	5-Hydroxymethlcytosine
1-Methylhypoxanthine	2-Thiocytosine
Xanthine	4-Acetylcytosine
6-Aminoacyladenine	
7-(4,5-*cis*-Dihydroxyl-l-clyclopenten-3-ylaminomethyl-7-deazaguanosine(Q)	

Source: From Ref. 3.

Complementary or Copy DNA (cDNA)

A DNA strand can be synthesized from an RNA template, in other words, the reverse of conventional transcription. This is accomplished by means of an enzyme, called transcriptase (RT), commonly found in RNA tumor viruses. A single-stranded DNA copy, termed "copy DNA" is obtained from a messenger RNA (mRNA). A double-stranded cDNA (ds = cDNA) can then be created by replication of the cDNA with DNA polymerase or reverse transcriptase after the original mRNA is removed by alkali degradation. In eukaryotes the mRNA represents a greatly condensed sequence, because the introns present in the DNA gene and in the original precursor RNA have been excised following transcription. Consequently, a cDNA will give only the exon sequences originally present in the DNA, but not the original complete DNA sequence as it exists in the gene.

The conversion of mRNA to cDNA was reviewed and reported in detail by Efstratiadis and Villa-Komaroff (21), covering methodology in general, creation of first strand, synthesis of second strand and construction of hybrid plasmids. The earlier publication of Maniatis et al. (22) serves as the basis of much of the report. The best characterized reverse transcriptase is AMV-RT, the enzyme isolated from avian myeloblastosis virus. The experimental conditions for cDNA synthesis are reported in Refs. 23 and 24. Some of the essential points stressed by the authors are discussed in the following paragraphs.

First Strand Synthesis: Control experiments should be conducted to ascertain the purity of the enzyme RT, especially for presence of RNase. Correct composition of the enzyme storage buffer is essential and final reaction pH should be 8.3. Concentration of mono- and divalent cations are critical, especially that of Mg^{2+}. Also, concentration of deoxynulceotide triphosphates should be greater than 50 μm.

Second Strand Synthesis: The formation of the cDNA second strand, the complementary structure, can be achieved with RT, Pol I, or T-4 polymerase. Temperature of reactions mixture should be kept at 15°C and the presence of KCl (but not NaCl) at a concentration of about 100 m*M* promotes strand formation. Frequently, hairpin formation is encountered and the loop can be opened with Sl nuclease.

Sequence of Events

The conversion of messenger RNA to complementary DNA involves some interesting intermediate steps. Normally, eukaryotic mRNA occurs in a poly-adenylated structure at the 3′ terminus. It is well known that the presence of a primer is essential in the reverse transcription process. Therefore, cDNA synthesis is achieved by adding oligo (dT) segments to the reaction mixture. These poly (dT)s hybridize with poly (A) structure of the mRNA and the hybrid sequence acts as a primer so that the cDNA growth begins at the 3′ end of the mRNA.

As shown in Figure 9 the use of reverse transcriptase plus (dT)s results in cDNA formation, where the (dT)s attach at the 5′ end. The cDNA strand tends to loop and form a so-called hairpin structure. As the reaction proceeds, the cDNA strand completely doubles back on itself till it reaches the (dT)-(dA) base pairing stage. Conventionally, the loop is opened by using Sl nuclease and the final product is a double-stranded cDNA, an exact copy of the original mRNA.

CHEMICAL SYNTHESIS OF DNA

Molecular cloning requires only microgram quantities of DNA. Once, the desired clone has been prepared, a continuous supply of DNA fragments is available. Thus, synthesized DNA has become very useful for recombinant DNA work.

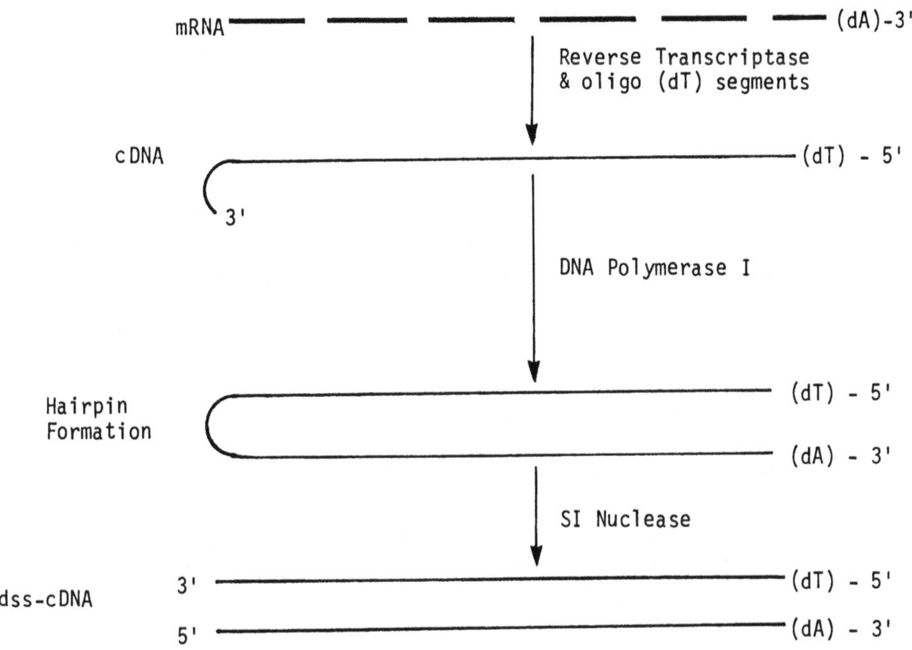

FIGURE 9 Reverse transcription. Starting with poly-tailed RNA, a complementary DNA (cDNA) is created by reverse transcription. Finally, the double-stranded DNA (dss-cDNA) is formed. It represents a DNA copy without the introns which occur in the original native DNA.

Synthetic applications include the use of linkers and primers (see Chap. 7), as well as synthesis, isolation, and alteration of genes (25).

There are several chemical synthesis methods which are utilized interchangeably, depending on the preference of the investigator. The methods are:

Commonly called	Chemical category
Diester method	Phosphodiester method
Triester method	Phosphate-triester method
	Phosphite-triester method
Enzymatic method	Polynucleotide phosphorylate method

Details of the various methods are given elsewhere (26-30).

Some essential comments follow.

Diester Method: A suitably protected mononucleotide and a similarly protected nucleoside are joined to form a di-dN containing a phosphodiester bond. The well-known procedure is diagrammed in Figure 10. Although this method is very time consuming, requiring hours or days for one addition, it apparently is still practiced by some investigators.

Triester Method: The *phosphate*-triester method uses an extra protecting group on the phosphate radicals of the reactants. This protecting group is usually a chlorophenyl group, which makes the N's and poly-N intermediates soluble in organic solvents. Purification can then be done in chloroform solutions. However, a more recent and preferred method uses coupling to a fixed bed matrix, silica gel, or polymeric beads. Figure 11 illustrates the skeletonized procedure.

Phosphite Triester Method: A *phosphite* nucleotide molecule is reacted with the nucleoside. Figure 11b shows the reaction scheme—note protecting groups. Oxidation yields the phosphate compound. Again solid support synthesis seems to be the preferred method of operation.

Phosphoramidite Method: This solid support method has gained widespread acceptance. It was described and illustrated by J. A. Smith (31) in a comprehensive treatise on the preparation of "short fragments of synthetic single-stranded DNA (15-20 bases) to serve as probes, primers, linkers, synthetic genes, and gene modifiers." (See Fig. 13a for a display and explanation of the procedure.)

Enzymatic Method: This is also the polynucleotide phosphorylase (pNPh) method. With proper control the pNPh promotes predominantly the addition of a single nucleotide (P-S-B) to a short oligodeoxy nucleotide. At least a trimer is required to start the reaction. The trimer must be obtained by another method. The desired single adduct can be obtained by chromatographic purification.

The stepwise addition by attaching the initial nucleotide to a solid support material is the basis of "gene machines," that is, automatic DNA synthesizers.

Synthesis (Gene) Machines

"Gene machines" represent a comprehensive assembly of instruments which are automated to produce specified short sequences of single-strand DNA (length of oligonucleotides up to about 40 bases).

One version, manufactured by Vega Biochemicals (26), uses packed columns, where nucleosides are "bound" to resin beads. The resins are described as being silica gel, and polystyrene or polymorpholine types. Solution containing the desired nucleoside (base plus sugar phosphate) is added in stepwise synthesis to add the particular base, one at a time. Nucleosides are commercially available from a number of sources. The Vega Biochemicals' model used addition of polynucleotide-protecting groups such as phenyl group, Cl, and triazole. After dimer formation, such groups are removed from the chain ends, allowing addition of chain fragments.

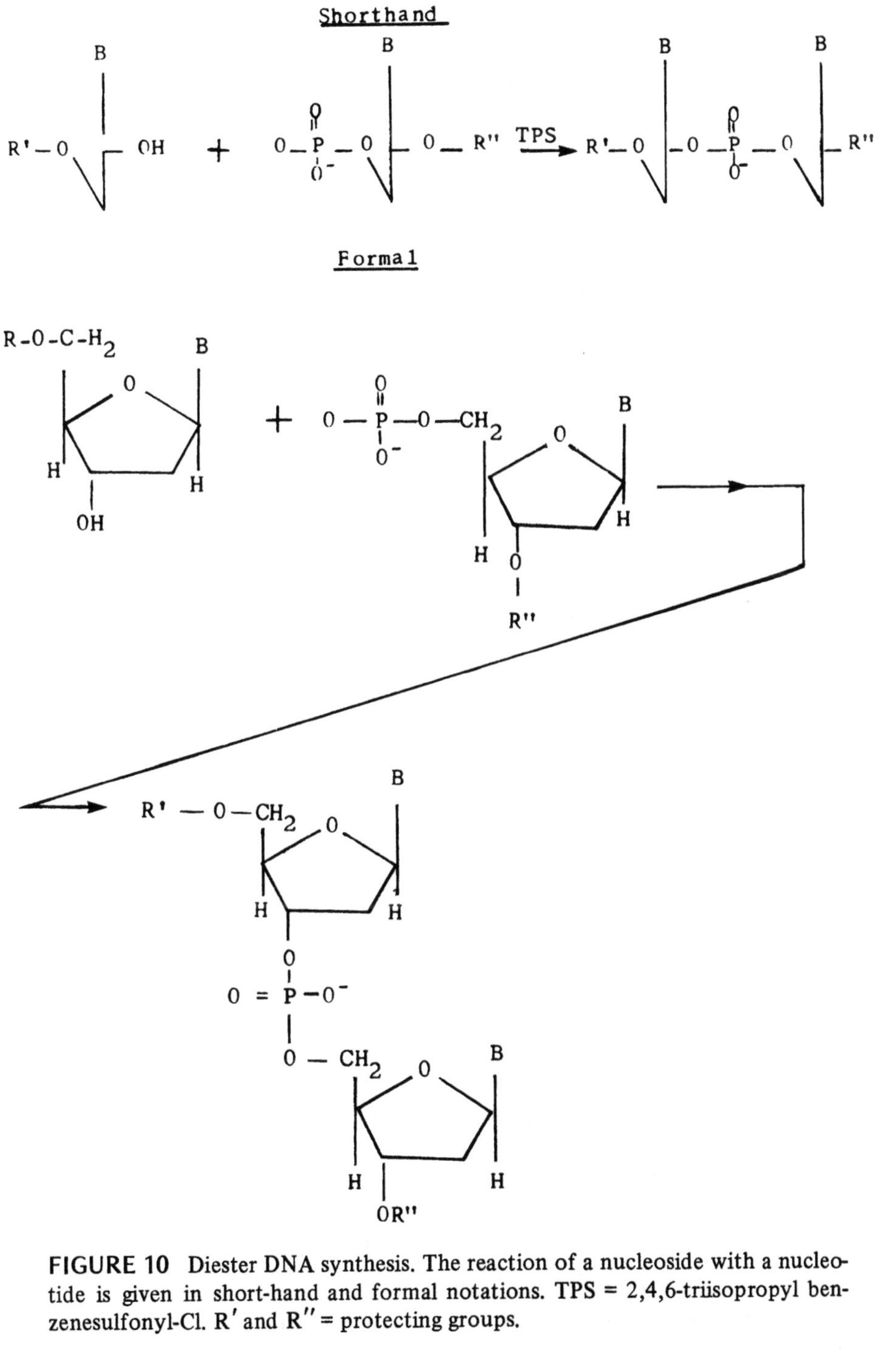

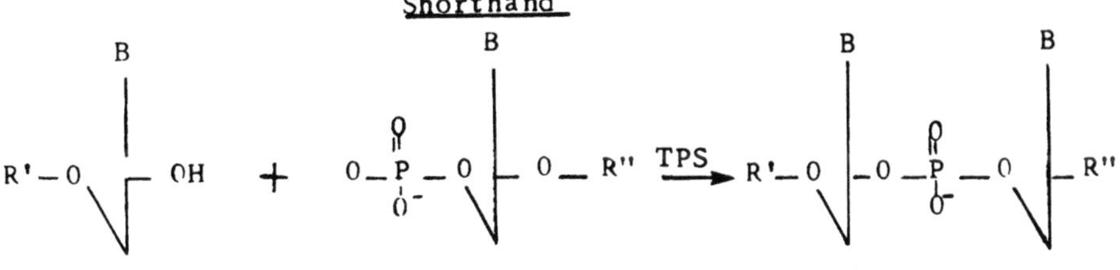

Formal

FIGURE 10 Diester DNA synthesis. The reaction of a nucleoside with a nucleotide is given in short-hand and formal notations. TPS = 2,4,6-triisopropyl benzenesulfonyl-Cl. R′ and R″ = protecting groups.

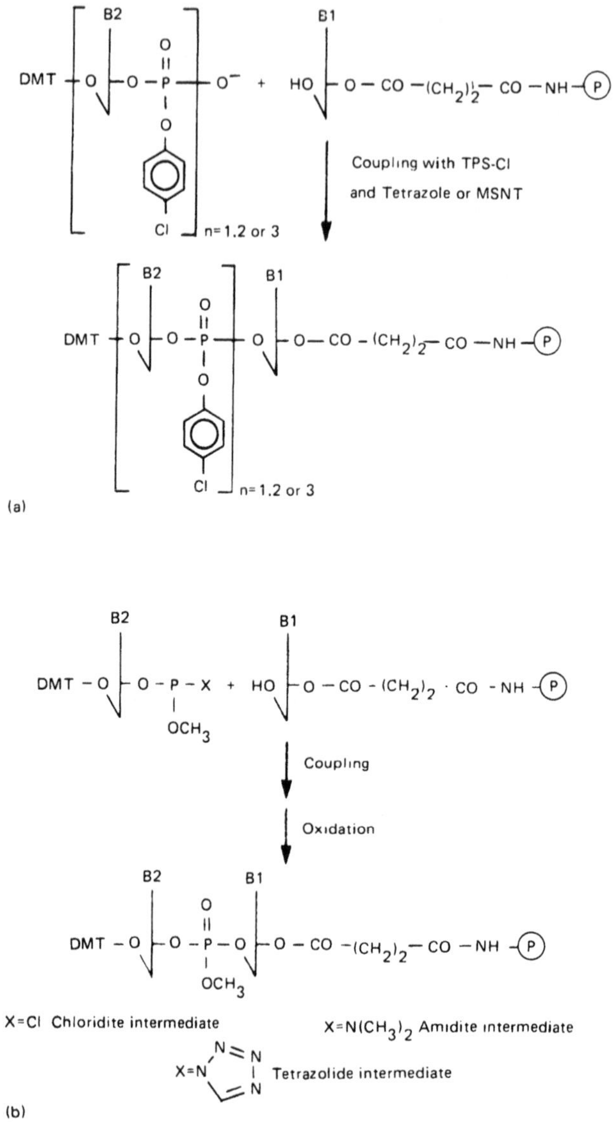

FIGURE 11 (a) Phosphate triester DNA synthesis. The reaction of a nucleotide sequence with a nucleoside is shown in short-hand notation. MSNT = mesitylene sulfonyl-3-nitro-1,2,4-triazole, TSP-Cl = 2,4,6-triisopropyl benzenesulfonyl-Cl, DMT = 4,4-dimethyoxytrityl group, P = silica gel, polystyrene polymer, or polydimethyl acrylate polymer, when fixed-bed synthesis is used. (Courtesy BIOSEARCH) (30). (b) Phosphite triester DNA synthesis. A phosphite nucleotide derivative is reacted with a nucleoside. Subsequent oxidation gives the normal phosphate compound. P = polymer matrix as in 3.11a. (Courtesy BIOSEARCH) (30).

The manufacture of these gene machines is a rapidly growing business. The procedures used in the instruments seem to follow a basic pattern. As such instrumentation is obviously very useful, it is worthwhile to present a somewhat detailed description of one type. The report on the BIO LOGICALS machine is rather more detailed than others and therefore the description by Alvarado-Urbina et al. will be used (27,29).

All solid support systems (28-30,32) use polymeric resins or silica gel. The latter is apparently preferred by most investigators and manufacturers, as the SiO_2 matrix has a much greater dimensional stability in exposure to organic solvents. The first preparative step is usually to create a silane linkage between the –SiOH radicals and an aliphatic silane. This procedure is commonly called "derivatizing." Details reported by Alvarado-Urbina are given in condensed form. Procedurally, the phosphite triester method is described.

The preparation of the solid support system is diagrammed in Figure 12. The significance of the lettering is as follows: S1 is silica gel activated by refluxing with HCl; further refluxing with the indicated silane in toluene gives a functionalized resin S2. Conversion of the amino group leads to stage S3; a protected nucleoside is then condensed to S3 using dicyclohexylcarbodiimide* as the condensing agent in pyridine; if diethylaminopyridine is the catalyst, the resin S4 will contain about 0.1 millimole of nucleotide per gram of resin; the final step is the removal of the dimethoxytrityle group (labeled DMT in the diagram), which is accomplished by digesting with mild acid giving the resin configuration S5. For use in the synthesizer apparatus the S5 product is used as packing in a small column. The amount is stated to be 200 mg. This is indicative of the scale of operation.

The container of the S5 resin at this point is inserted into the machine. From solution reservoirs the reagents are pumped through the resin sequentially as shown in the diagram of Figure 13. The individual steps are outlined: the active phosphorylating agent (1) converts S5 to a coupled phosphite S6, which is then oxidized to phosphate S7 with acqueous idone. Water removal by means of phenylisocyanate, chloroform flushing to remove excess reagent, DMT excision with mild acid, and a solvent wash, yield the final product S8, a nucleotide dimer. This completes the cycle.

As far as can be ascertained, the gene machines produced by the various manufacturers follow a common scheme. Variations consist in redundancies of equipment items, such as valves, and number of solution containers. Also, sequences of reactions and type of reactants may differ. Electronic circuitry to make procedures automatic are likely to be unique for the different models. Generally, the valves are operated by solenoids controlled by microprocessing

---

*More recently arylsulfonylhalides and arylsulfonyltetrazoles are being used.

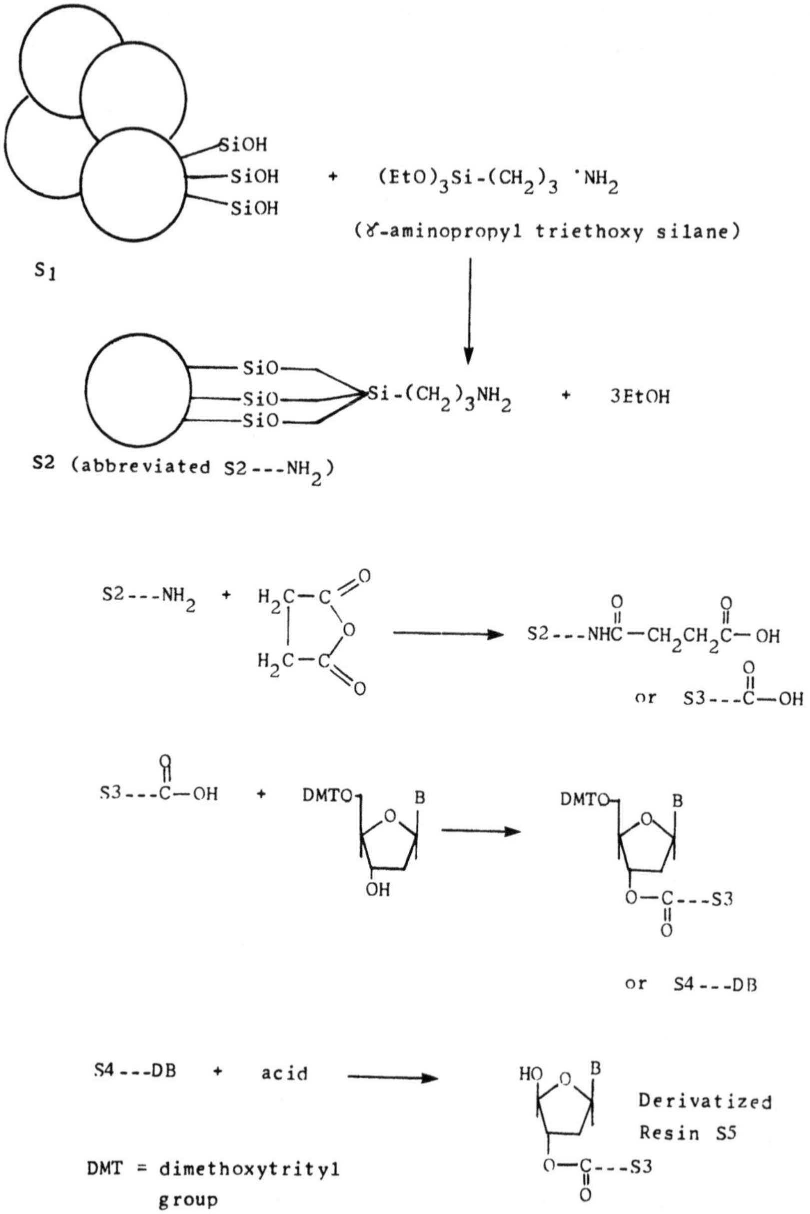

SiOH
SiOH + $(EtO)_3Si\text{-}(CH_2)_3 \cdot NH_2$
SiOH
(γ-aminopropyl triethoxy silane)
S_1

SiO
SiO $Si\text{-}(CH_2)_3NH_2$ + 3EtOH
SiO
S2 (abbreviated $S2\text{---}NH_2$)

$S2\text{---}NH_2$ + $H_2C{-}C(=O)$ / O / $H_2C{-}C(=O)$ ⟶ $S2\text{---}NHC(=O){-}CH_2CH_2C(=O){-}OH$

or $S3\text{---}C(=O){-}OH$

$S3\text{---}C(=O){-}OH$ + DMTO, B, O, OH ⟶ DMTO, B, O, $O{-}C(=O)\text{---}S3$

or S4---DB

S4---DB + acid ⟶ HO, O, B, $O{-}C(=O)\text{---}S3$ Derivatized Resin S5

DMT = dimethoxytrityl group

FIGURE 12 Steps in automatic DNA synthesis. Silica resin as the solid matrix is reacted in sequential steps to yield the derivatized state used in subsequent synthesis.

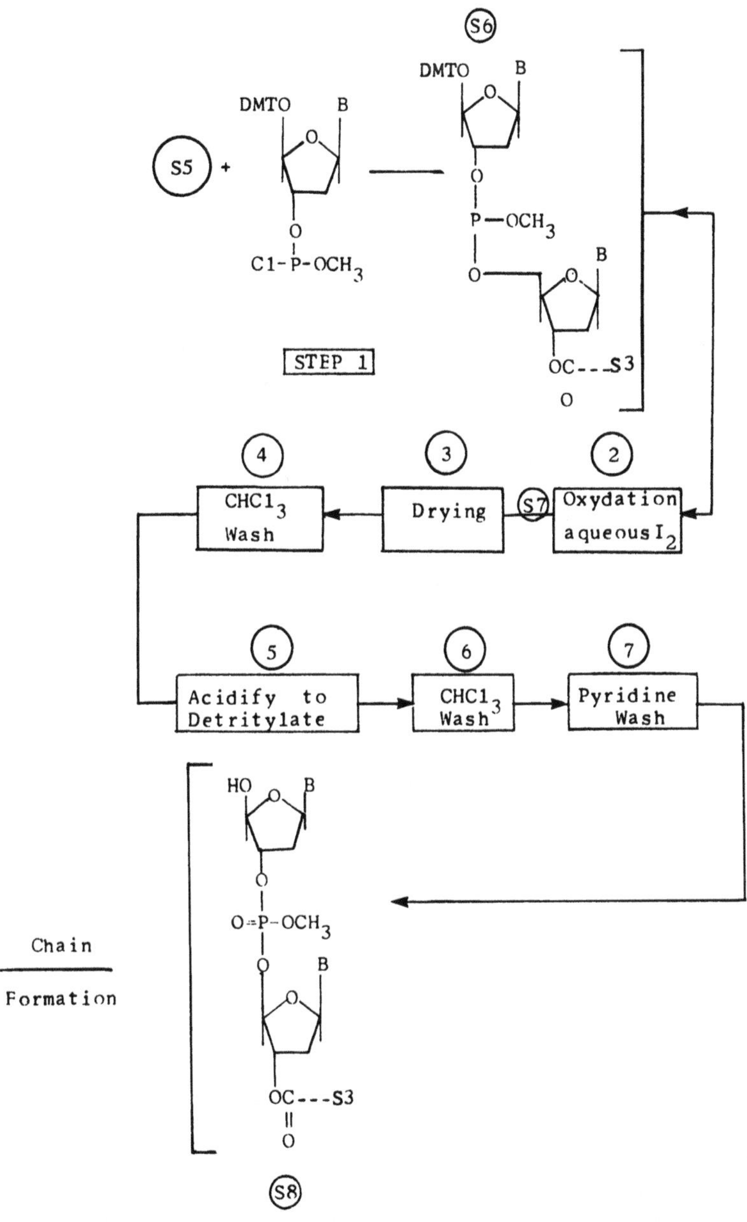

FIGURE 13a DNA synthesis in gene machine. The derivatized resin prepared as diagrammed in Figure 12, is used for nucleotide additions by the phosphite triester method (b) Phosphoramidite synthesis on solid support. Contained in figure. Reprinted from *American Laboratory*, see Ref. 31, page 16. Copyright 1983.

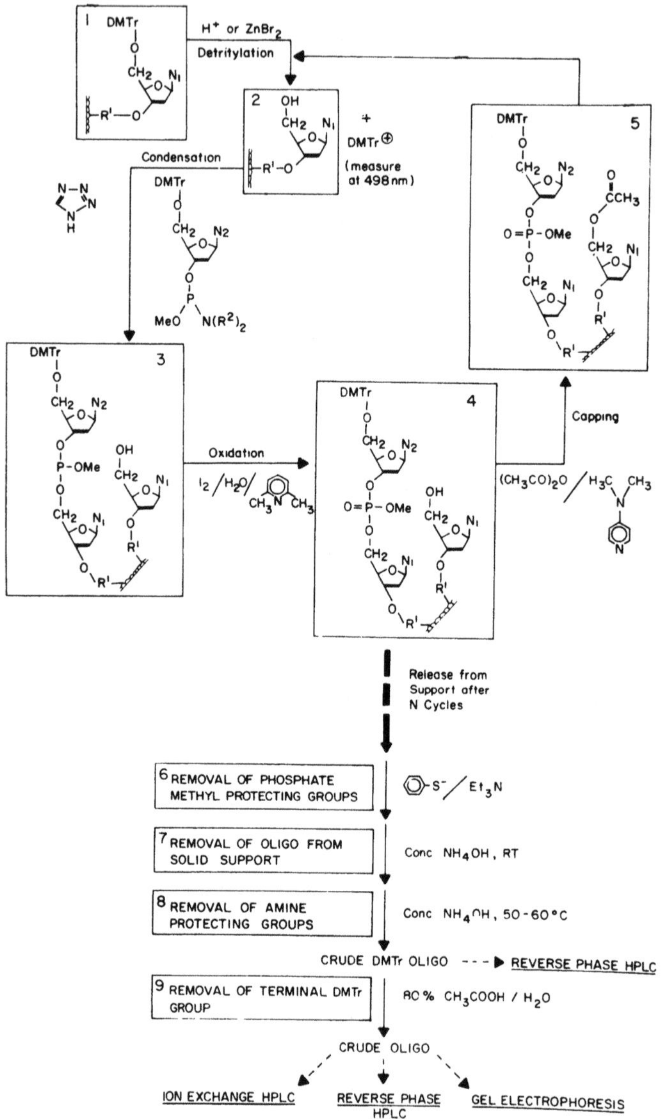

FIGURE 13b Phosphoramidite synthetic route: 1–2) Detritylation with trichloroacetic acid or zinc bromide removes the 5′-dimethoxytrityl (DMTr) from the support-linked base or oligodeoxyribonucleotide (oligo). The DMTr cation's brilliant orange color can be measured at 498 nm, and the absorption values can be used to calculate coupling efficiencies between different synthetic cycles. R^1

signals and the complete sequence operation is put into a computer. The flow diagram in Figure 14 is modeled from the description by Alvaredo-Urbina et al. (27), who delineated the functioning of the BIO LOGICAL's model. This machine has been redesigned for improved mechanical reliability and flexibility.

A gene machine procedure was described earlier by Matteuci and Caruthers (28). It uses a polymer support consisting of chromatography type silica gel to which *d*-nucleosides are bonded through a carboxylic functional group. The method and procedural steps are presented in great detail in the publication.

A popularized version of the gene machine story appeared in the July/August issue of *Science 81* (27). There the Vega Biochemical and the BIO LOGICAL's machines are described and prices are quoted at under $30,000.* Also discussed is work at Cal Tech directed toward automatic sequencing of proteins using liquid chromatography for amino acid identification with mass spectrometry contemplated for future refinement.

A well-designed and competitively priced synthesizer is manufactured by Biosearch (30). The scheme of chemical synthesis is analogous to that described by Alvaredo-Urbina et al. (27), that is, the use of a solid matrix. The Biosearch

---

is an organic spacer arm separating the oligo from the surface of the solid support. N_1 is a purine or pyrimidine base. 2–3) Condensation of the protected phosphoramidite with the support-linked base in the presence of tetrazole. R^2 is a phosphoryl protecting group (e.g., methyl, ethyl, isopropyl). N_2 is a purine or pyrimidine base. 3–4) Oxidation of the reactive phosphite to a stable triester by aqueous iodine containing lutidine. 4–5) Capping of unreacted 5′-OH groups by acetylation with acetic anhydride in the presence of dimethylaminopyridine. 5–2) Initiation of next coupling cycle. Coupling cycles are repeated until an oligo of the desired sequence is completed. 6) Removal of the phosphoryl methyl protecting group by thiophenoxide ion formed from thiophenol in the presence of triethylamine. 7) Removal of oligo and contaminants by base (ammonia) hydrolysis at the succinate linkage. 8) Removal of amine protecting groups from deoxyadenosine, deoxyguanosine, and deoxycytosine by base (ammonia) hydrolyisis at 50–60°C. 9) Removal of the terminal 5′-DMTr group by concentrated acetic acid. An oligo containing a 5′-DMTr group is more hydrophobic than oligos in the mixture with free 5′-OH groups and are readily separated by reversed-phase (RP) high performance liquid chromatography (HPLC). Ion exchange HPLC, RP–HPLC, or gel electrophoresis can be used to purify an oligo from a mixture of truncated (i.e., oligos with shortened sequences), nonsense (i.e., oligos with jumbled sequences), and partially protected oligos (i.e., oligos with protecting groups remaining).

---

*Such quotes are likely to be obsolete in the foreseeable future.

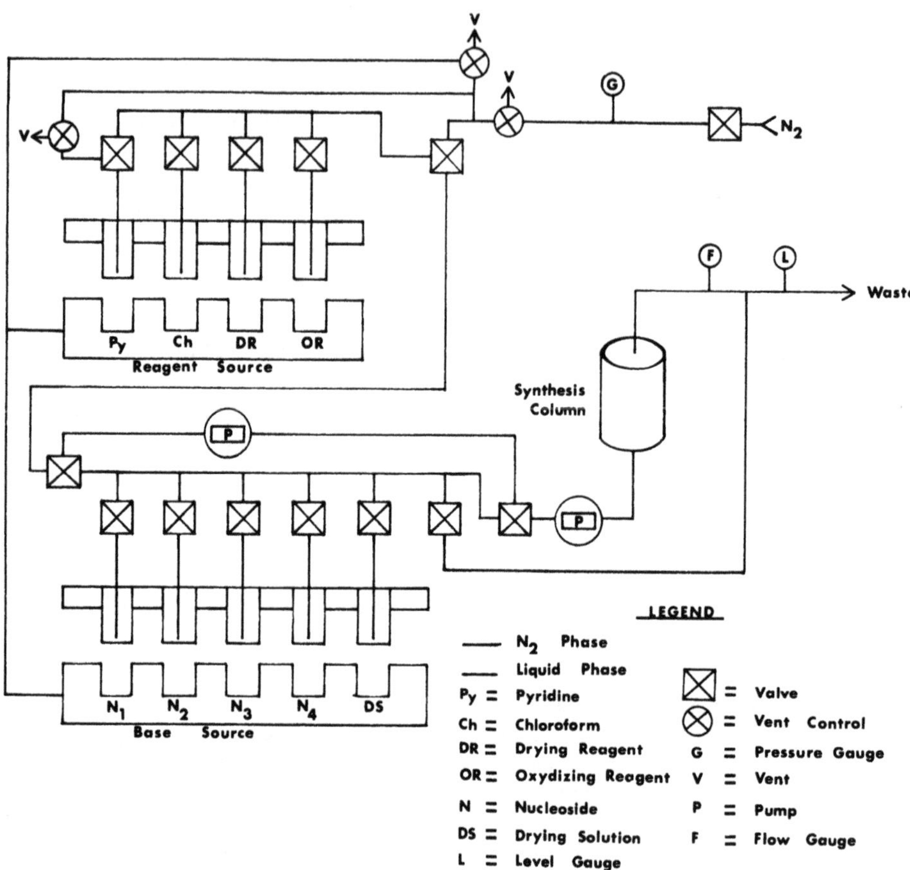

FIGURE 14 Automated DNA synthesis. Generalized flow diagram for sequential steps in a gene machine. The casettes holding reagents and nucleosides are arranged for snapping into the stationary equipment after containing wells are filled with the desired chemicals. Nitrogen pressuring may be used for fluid movement and blowing down of lines. (Assistance by Dr. Alvaredo-Urbina is gratefully acknowledged.)

unit, however, uses a high-pressure column to conduct synthesis. The instrument is marketed as the SAM DNA synthesizer. Extension to polypeptide synthesis, polysaccharides, and other chemical reactions is envisioned. The SAM machine is programmed to perform both the phosphate-triester and the phosphite-triester synthesis.

As one might expect, the manufacturers of gene machines also offer a wide selection of reagents and related biochemicals.

REFERENCES

1. Watson, JD, Crick, FHC. A structure for deoxyribonucleic acid. *Nature* 171: 737–738 (1953).
2. Watson, JD. (1980). *The Double Helix*. Norton Critical Editions in the History of Ideas, Norton.
3. Adams, RLP et al. (1981). *The Biochemistry of the Nucleic Acids*, 9th Ed. Chapman and Hall, New York.
4. Neidler, S (Ed.). (1981). *Topics in Nucleic Acid Structure*. John Wiley and Sons, New York.
5. Watson, JD, Tooze, J. (1981). *The DNA Story*. W. H. Freeman & Co., San Francisco.
6. Watson, JD. (1976). *Molecular Biology of the Gene*, 3rd Ed. W. A. Benjamin, Inc.
7. Keeton, W. (1980). *Biological Science*, 3rd Ed. W. W. Norton & Co., New York.
8. Orgel, LE, Crick, FHC. Selfish DNA: The ultimate parasite. *Nature* 284: 604–607 (1980).
9. Lewin, R. Can genes jump between eukaryotic species? *Science* 217: 42–43 (1982).
10. Wang, AH-J et al. Molecular structure of a left-handed double helical DNA fragment at atomic resolution. *Nature* 282: 680–686 (1979).
11. Norman, C. *Science* 214: 1108–1110 (1981).
12. Fisher, LM. DNA supercoiling by DNA-gyrase. *Nature* 294: 607–608 (1981).
13. Wang, JC. DNA topoisomerases. *Sci. Am.* 247: 94-109 (1982).
14. Kornberg, RD Klug, A. The nucleosome. *Sci. Am.* 244: 52–64 (1981).
15. Kornberg, A. *DNA Replication*. W. H. Freeman & Co., San Francisco.
16. Holley, RW. The nucleotide sequence of a nucleic acid. *Sci. Am.* 214: 30–39 (1966).
17. Wisseroth, K. Life span and carcinogenesis—the chemical aspects (in German). *Chemiker Zeitung* 107: 1–13 (1983).
18. Benzinger, TH. Thermodynamics of living matter: physical foundations of biology. *Am. J. Physiol.* 244(L): R743–750 (1983).
19. Freifelder, D. (1982). *Physical biochemistry—Applications to Biochemistry and Molecular Biology*, 2nd Ed. W. H. Freeman and Co., San Francisco.

20. Hawker, LE, and Linton, AH. (1979). *Micro-organisms*, 2nd Ed. University Park Press, Baltimore.
21. Efstratiadis, A, Villa-Komaroff, L. (1979). Cloning of double-stranded cDNA. In *Genetic Engineering*, Setlow, JK, Hollaender, A (Eds), Plenum Press, New York.
22. Maniatis, T, et al, *Cell 7*, 279–288 (1976); *Cell 8*, 163–182 (1976).
23. Efstratiadis, A. et al. (1976). In *Methods of Molecular Biology*. Marcel Dekker, New York, p. 1.
24. *Molecular Mechanisms in the Control of Gene Expressions, 5* (1976), Academic Press, New York, p. 513.
25. Itakura, K, Riggs, AD. Chemical DNA synthesis and recombinant DNA studies. *Science* 209: 1402–1405 (1980).
26. Vega Biochemicals Catalog. P.O. Box 11648, Tuscan, AZ 85734.
27. Alvarado-Urbina, G et al. Automated synthesis of gene fragments. *Science* 214: 270–274 (1981).
27a. Menosky, JA. The gene machine. *Science 81* 2(6): 38–41 (1981).
28. Matteucci, MD, Caruthers, MH. Synthesis of deoxyoligonucleotides on a polymer support. *J. Am. Chem. Soc.* 103 (11):3185–3191 (1981).
29. BIO LOGICALS Catalogue. DNA/RNA Synthesizer, BIO LOGICALS, 7 Hinton Ave., North, Ottawa, Canada KIY4PI.
30. BIOSEARCH Catalog, SAM Synthesizer, BIOSEARCH, 1281-F Andersen Drive, San Rafael, CA 94901.
31. Smith, JA. Automated solid phase oligodeoxyribo-nucleotide synthesis. *Am. Biotechnol. Lab.* 1(1): 15–24 (1983).
32. BACHEM, Inc. Nucleotides Bulletin, Torrance, CA.

4
Protein Synthesis

PROTEINS

Proteins are polymers made up of linear sequences of amino acids. There are 20 basic amino acids (see Table 1) which are considered the essential building blocks for the great multitude of possible protein combinations. All amino acids contain nitrogen atoms, therefore the proteins represent a class of nitrogenous compounds. Linkage of the amino acids is by means of a peptide bond (Fig. 1) and so the polymer is a polypeptide chain. Proteins constitute as much as two-thirds of the dry weight of cells and their molecular weights vary from fairly small structures, as for instance the insulin molecule, to very large configurations having molecular weights in excess of one million.

Table 2 illustrates the variation of the amino acid content of proteins. Of the three proteins illustrated, two are enzymes and one is for the hormone insulin, which as shown is one of the smaller proteins known with only 51 amino acids.

A most important development in molecular biology was the establishment of the *genetic code* (1). When it was realized that proteins were made up of only 20 amino acids, it was possible to relate these to the four nucleotide bases which were thought to be responsible for directing protein synthesis from their respective sequences in DNA and RNA. It became evident that a sequence of three bases was responsible for coding one amino acid in a polypeptide chain. Then, with four bases such a scheme would allow for $4^3 = 64$ combinations to code for the 20 amino acids.

A triplet of bases in the mRNA is called a *codon*. It codes for a specific amino acid, and a specie of low molecular weight RNA, the tRNA, accepts the amino acid and transfers it to the polypeptide chain. The unique cloverleaf structure of tRNA was shown in Chapter 3, Figure 8.

TABLE 1 Amino Acids

Alanine (Ala)	Glycine (Gly)	Proline (Pro)
Arginine (Arg)	Histidine (His)	Serine (Ser)
Asparagine (Asp-NH_2 ,Asn)	Isoleucine (Ile)	Threonine (Thr)
Aspartic acid (Asp)	Leucine (Leu)	Tryptophan (Trp)
Cysteine (Cys)	Lysine (Lys)	Tyrosine (Tyr)
Glutamic acid (Glu)	Methionine (Met)	Valine (Val)
Glutamine (Glu-NH_2 ,Gin)	Phenylalanine (Phe)	

Source: From Refs. 2a and 2b.

Genetic Code

The amino acids (20 in number) involved in protein synthesis are listed in Table 1. The codons (genetic code for RNA) that is, triplets of mRNA combinations, which direct the formation of a specific protein sequence are shown in Table 3. Figure 2a shows the coding arrangement, where base triplets in the mRNA direct the amino acid selection. Also shown in Figure 2b is the coding directive given by the DNA to form the RNA strand by the principle of base pairing. The lower figure indicates the mRNA strand formation from one of the DNA strands.

The amino acid chain of an enzyme is shown schematically by Figure 3 where disulfide bridges between sections of the polymer chain are indicated (–S–S–). Such bonds are located at cysteine positions. Enzymes and many hormones are proteins.

Polypeptide Formation

The polypeptide chain is created by the interaction of amino acid tRNA with mRNA. The process itself takes place on the ribosomes. There, successive codons are read in an ordered sequence and the amino acids are linked stepwise to form the protein chain. Direction of growth is from the amino terminus to the carboxyl end, with mRA directive in the 5′ to 3′ direction. The rates of protein synthesis are of the same order of magnitude for prokaryotes and eukaryotes, namely in the range of 8 to 15 amino acids per second.

$$R-\underset{NH_2}{\overset{H}{C}}-\overset{O}{\overset{\|}{C}}-OH \;+\; H-\overset{H}{N}-\underset{COOH}{\overset{H}{C}}-R' \longrightarrow R-\underset{NH_2}{\overset{H}{C}}-\overset{O}{\overset{\|}{C}}-\overset{H}{N}-\underset{COOH}{\overset{H}{C}}-R' \;+\; H_2O$$

AMINO ACID I AMINO ACID II DIPEPTIDE WATER

FIGURE 1 Linkage of amino acids by means of a peptide bond.

TABLE 2 Amino Acid Composition of Three Proteins

Amino acid	*E. coli* β-galactosidase Mole percent	Number per chain	*E. coli* tryptophan synthetase (A protein) Mole percent	Number per chain	Bovine insulin Mole percent	Number per chain
Ala	8.0	93	15.0	40	6.0	3
Arg	6.3	74	4.1	11	1.9	1
Asn	10.5	123	8.2	22	6.0	3
Asp					–	0
Cys	1.6	19	1.9	5	11.8	6
Gln	12.1	142	10.9	29	6.0	3
Glu					7.8	4
Gly	7.2	85	7.1	19	7.8	4
His	3.1	36	1.5	4	3.7	2
Ile	4.1	48	7.1	19	1.9	1
Leu	9.4	110	10.1	27	11.7	6
Lys	2.5	29	4.9	13	2.0	1
Met	2.1	24	1.1	3	–	0
Phe	3.8	45	4.5	12	6.0	3
Pro	5.7	67	7.1	19	1.9	1
Ser	5.7	67	4.1	11	6.0	3
Thr	5.5	65	3.3	9	2.0	1
Trp	3.0	35	–	0	–	0
Tyr	3.1	36	2.7	7	7.7	4
Val	6.3	75	6.4	17	9.8	5
Total	100.0	1173	100.0	267	100.0	51

Source: From Ref. 1.

TABLE 3 Code Table for the Base Triplets of mRNA

First Position (5′ end)	Second Position: U		C		A		G		Third Position (3′ end)
U	UUU	Phe	UCU	Ser	UAU	Tyr	UGU	Cys	U
	UUC	Phe	UCC	Ser	UAC	Tyr	UGC	Cys	C
	UUA	Leu	UCA	Ser	UAA	Ochre	UGA	Umber	A
	UUG	Leu	UCG	Ser	UAG	Amber	UGG	Trp	G
C	CUU	Leu	CCU	Pro	CAU	His	CGU	Arg	U
	CUC	Leu	CCC	Pro	CAC	His	CGC	Arg	C
	CUA	Leu	CCA	Pro	CAA	Gln	CGA	Arg	A
	CUG	Leu	CCG	Pro	CAG	Gln	CGG	Arg	G
A	AUU	Ile	ACU	Thr	AAU	Asn	AGU	Ser	U
	AUC	Ile	ACC	Thr	AAC	Asn	AGC	Ser	C
	AUA	Ile	ACA	Thr	AAA	Lys	AGA	Arg	A
	AUG	Met	ACG	Thr	AAG	Lys	AGG	Arg	G
G	GUU	Val	GCU	Ala	GAU	Asp	GGU	Gly	U
	GUC	Val	GCC	Ala	GAC	Asp	GGC	Gly	C
	GUA	Val	GCA	Ala	GAA	Glu	GGA	Gly	A
	GUG	Val	GCG	Ala	GAG	Glu	GGG	Gly	G

The codons AUG or occasionally GUG specify initiation of peptide chain formation. The codons UAG, UAA, or UGA specify termination of peptide chain growth. The termination codon UAG is sometimes called Amber. The termination codon UAA is sometimes called Ochre. The termination codon UGA is sometimes called Opal or Umber. More recent code tabulation avoid using amber, ochre, opal, and umber. In their place the expression "Term" is used, signifying "termination."

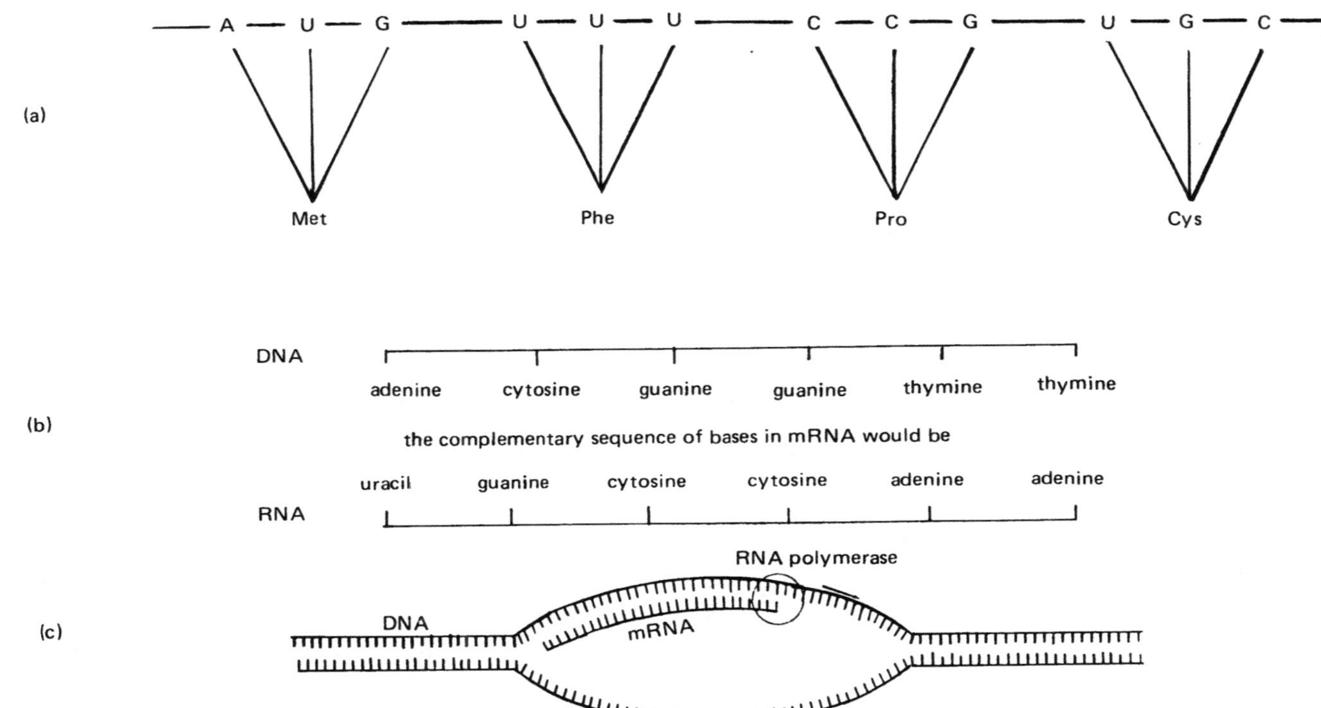

FIGURE 2 Coding Scheme. (a) Triplet codons on MRNA code for amino acids. (b) DNA strand codes by base-pairing to form RNA. (c) ds-DNA starts unwinding and the coding strand forms mRNA. (Courtesy McGraw-Hill, Ref. 3).

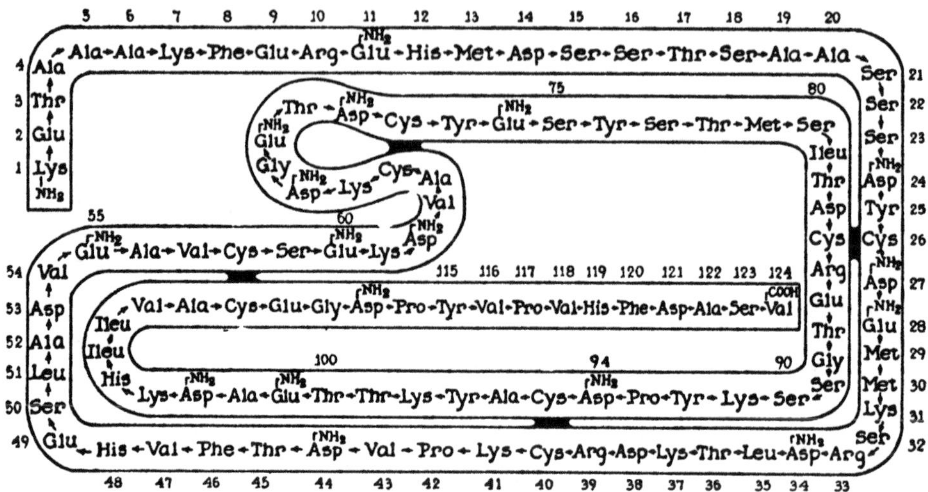

FIGURE 3 Amino Acid sequence in a protein chain. The amino acid sequence of the enzyme ribonuclease. The shaded areas between cysteines represent disulfide bridges. This illustration is diagrammatic; the polypeptide chain is actually folded to give a complex three-dimensional configuration. (From Ref. 3.)

Extent of Coverage

In every detail, the process of protein formation is a highly complex succession of events. Adams et al (4) characterize the present status as still being unresolved in all of its reaction sequences. The interlocking reactions present a formidable task of complete interpretation.

Therefore, the question is how much coverage should be given to the subject in the present text. In line with the practice of most general biology texts, it is appropriate to present a highly simplified version and then discuss essential differences in prokaryotic and eukaryotic polypeptide synthesis.

Sequence of Protein Formation

In prokaryotes and eukaryotes alike, the complete process involves the following series of steps:

Replication: the synthesis of DNA in the cell
Transcription: the synthesis of RNA

(Replication and transcription have been detailed in Chap. 3.)
Translation: the multiplicity of reactions resulting in protein chain initiation, elongation, and chain termination.

The overall sequence in a simplified form, would be as follows:

1. DNA in the cell codes an mRNA. In the eukaryote this takes place in the nucleus; there is an appreciable difference in the behavior of prokaryotes and eukaryotes (see later discussion).
2. The mRNA, then associates with a free ribosome in the cell cytoplasm.
3. Simultaneously to steps 1 and 2, strands of tRNA, combine with their specific amino acid also present in the cytoplasm.
4. The mRNA from step 2 was formed to contain specific codon sequences which interact with so-called corresponding anticodons, located on the "tRNA-amino acids" complex from step 3. Start codon is usually AUG for methionine (MET). Stop codons are UAA, UAG, and UGA (Table 3).
5. Thus, progressively an amino acid complex is formed by means of peptide bonding, the *translation* process.
6. When the complex is finished, a codon signals the end of polymerization and the protein molecule, a polypeptide is freed to go to its specific location in the organism as dictated by the original gene command from the DNA.

The ribosome plays an important intermediate part in protein formation. The various steps involved have been presented in publications by Gassen (5), Lake (6), and Hopwood (7). The interlocking processes were illustrated in a systematic and lucid manner. Figure 4 represents a composite of Gassen's illustrations. Thus, at upper left, the initiation procedure is shown, leading to the central figure which illustrates protein elongation. The final step, termination of protein synthesis appears at upper right. It is important to realize that the *same ribosome* is involved in the sequential steps: the initiation step starts the protein synthesis on the ribosome; at its conclusion the ribosome reaction shifts over to elongation; when the stop signal comes along another shift brings on the termination reaction. One can visualize the coding mechanisms as a taped message on a moving ribbon, or in other words, a tickertape.

Chain Growth

The addition of an amino acid to the growing peptide chain is an energy-consuming process. So, the organism provides the amino acid molecule in an intermediate activated form as an amino acid ester. It is called an *amino-acyl-tRNA*. The structure for an adenine ribose complex is shown in Figure 5.

The tRNA structure in this cloverleaf form was illustrated previously. The binding of the amino acid to tRNA serves to activate its carboxyl group for the peptide formation. Also the tRNA tail recognizes the codons on the mRNA and so assures the proper placing of the amino acid in the protein sequence.

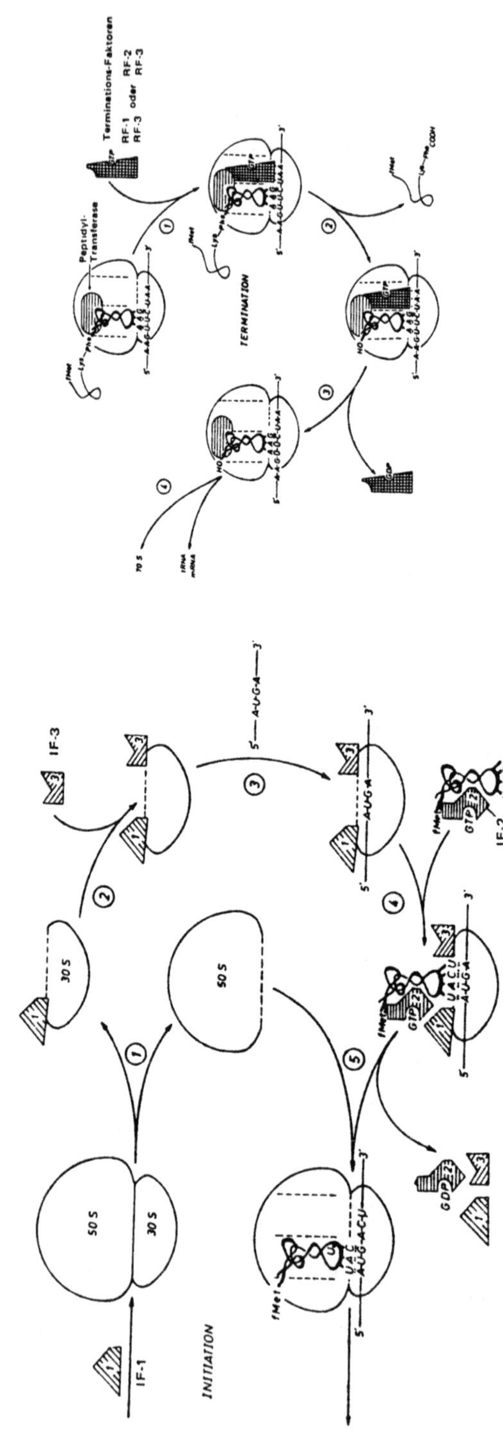
INITIATION
IF-1
50 S
30 S
IF-3
IF-2
GTP
GDP
TERMINATION
Peptidyl-Transferase
Terminations-Faktoren

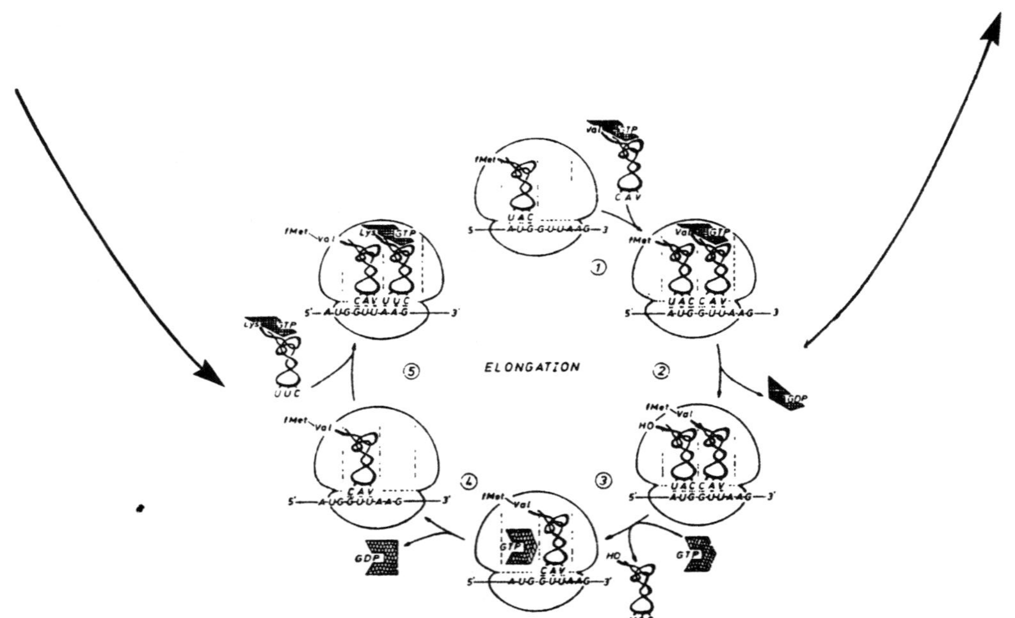

FIGURE 4 The ribosome protein synthesis. Initiation, upper left; elongation, lower center; termination, upper right. If 1, 2, and 3 are elongation factors, and RF 1, 2, and 3 are termination factors, then GTP = guanosine triphosphate, GPD = guanosine diphosphate, f-Met = *N*-formyl methionine, and V = modified uridine. (From Ref. 5.)

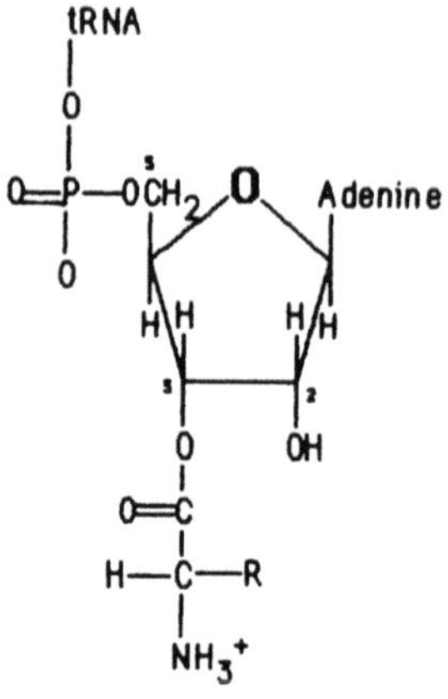

FIGURE 5 Adenine-ribose complex.

DIFFERENCES IN PROKARYOTIC AND EUKARYOTIC DNA AND RNA

Bacterial DNA-RNA

In bacteria and in bacterial viruses, the DNA, both chromosomal and extrachromosomal, are double-stranded DNA forming uncomplicated genomes. There is little, if any, "unexpressed" information (usually less than 10%) in the DNA. So, most of the DNA sequence is responsible for directing protein synthesis via RNA formation.

The important consequence of the relative simplicity of bacterial DNA structure is that in prokaryotes the structural gene is transcribed directly into messenger RNA, without intervening sequences which then translates into protein synthesis. This is illustrated in Figure 6a. In prokaryotes, translation of the messenger RNA begins before transcription is completed; this is called coupled transcription/translation.

In prokaryotes, specifically *Escherichia coli*, a particular tRNA carries a derivative of methionine at its 3′ chain terminus, namely *N*-formyl methionine Fig. 7). This compound is introduced at the amino terminal end of the polypeptide chain where the growth of the chain is initiated.

The formylated methionine attaches to the ribosomal aminoacyl site, signaling the start of protein formation. The complex is usually written as fMet-$tRNA_f^{Met}$ and its tRNA has a different nucleotide sequence than the tRNA, which inserts Met internally in the protein chain. The occurrence of the fMet compound is unique for prokaryotic RNA formation: it does not occur in eukaryotes. However, eukaryotes also have a unique tRNA, $tRNA_i^{Met}$, that functions only as the tRNA involved in initiation and does not function to add internal methionine residues.

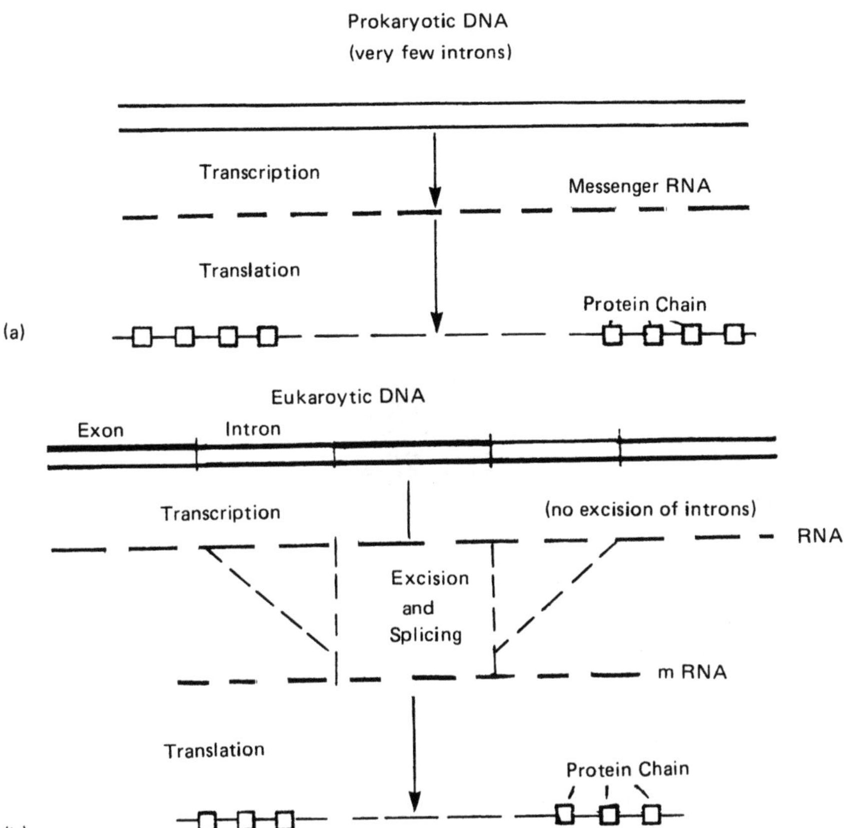

FIGURE 6 Expression of DNA. (a) Transcription of genetic information in a prokarytic cell (bacterium) is complete. So the RNA is a duplicate of the DNA, except for U in place of T. (b) Transcription of genetic information in a eukaryotic cell shows the intermediate step of excission of introns before the operative mRNA is produced.

```
  O H H O
  | | | |
H-C-N-C-C-O⁻
      |
      CH2
      |
      CH2
      |
      S
      |
      CH3
```

FIGURE 7 *N*-formyl methione.

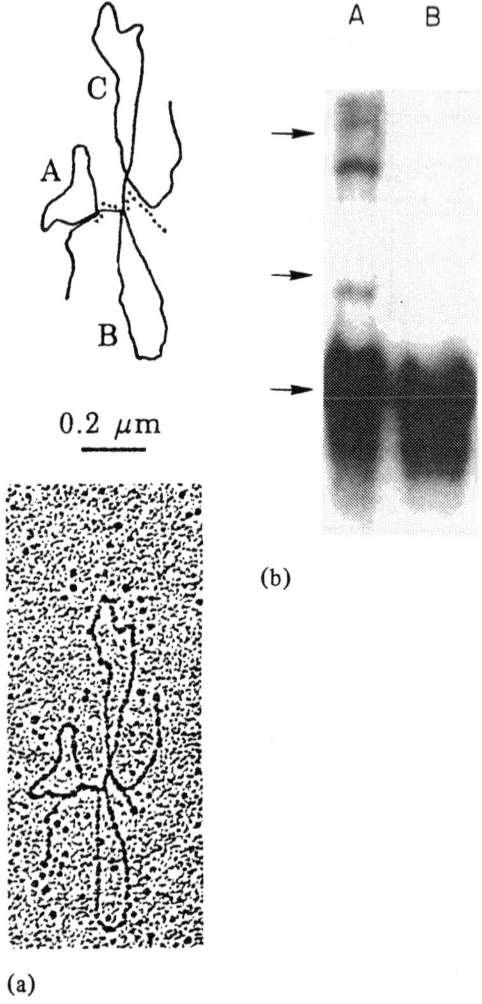

FIGURE 8 Intron excision from prolactin precursor RNA. (a) Intron loops in prolactin precursor are illustrated; the excision of the loops A, B, and C results in the formation of mRNA. (From Ref. 8.) (b) The before and after excision state is visualized in the picture of a gel electrophoresis autograph. (From Ref. 9).

Eukaryotic DNA-RNA

In eukaryotic organisms, the DNA contains an appreciable number of noncoding regions. Sometimes such regions intervene with coding regions and the term "split genes" is used to describe the condition. The noncoding sequences are called "introns" and the coding sequences are "exons." When DNA directs the formation of a primary RNA, all sequences are transcribed; the ensuing structure is a precursor mRNA. In a subsequent step, the intron transcripts are excised and the exon sequences are spliced together to create the mRNA. Figure 6b illustrates the "splicing" operation.

The appearance of "RNA-loops" in a prolactin precursor RNA is illustrated in Figure 8 a as an electron micrograph and a sketch of the RNA from the photograph. The loops marked *A, B,* and *C* represent intervening sequences (introns) which are being excised. Figure 8b illustrates the results of gel electrophoresis, with lane *A* before excision and lane *B* after excision of introns. The looping of the introns is shown diagrammatically in Figure 9.

Overall Sequence of Protein Synthesis

The overall process in the *prokaryotic cell,* that is, DNA to protein is illustrated diagramatically in Figure 10 showing the sequence: DNA coding strand *C* directing transcription to mRNA followed by triplet coding in translation to the protein chain.

A pictorial scheme of events taking place in a *eukaryotic cell* is sketched in Figure 10. The sequences, in outline, are as follows:

ds-DNA in the nucleus unwinds and codes for precursor;
m-RNA (pm-RNA); the DNA itself does not leave the nucleus;
pm-RNA undergoes intron excision to form m-RNA; excision step is not shown
m-RNA leaves the nucleus and moves to ribosome;
t-RNA forms complexes with amino acids: complex moves to ribosome;
Interaction at ribosome gives the properly coded protein chain;
Protein chain is released from ribosome.

An additional step occurs for proteins that are excreted from the cell (exocytosis).

A most instructive source on proteins was published by Dickerson and Geis (10), which presents in great deal and with numerous lucid illustrations, the structure and action of proteins.

Eukaryotic Gene

with

Intervening Sequences

Transcription to pre-mRNA

Formation of Loop Structures

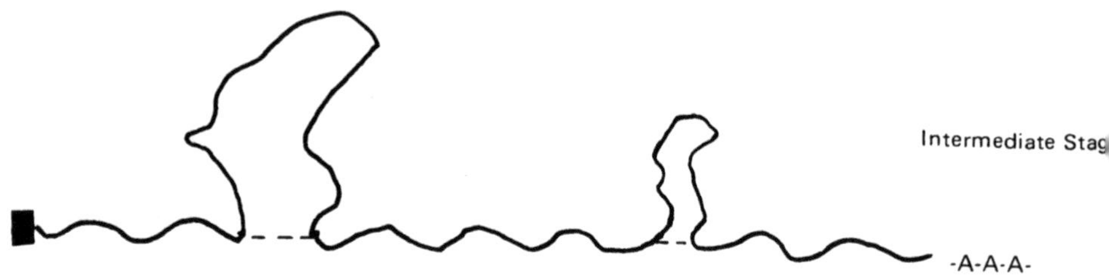

Excision and Splicing

Condensed Gene in RNA

FIGURE 9 Maturation of mRNA. The sequence involved in "looping" out the introns from a eukaryotic gene is ilustrated. The significant portion of the figure is the looped configurations. The formation of such loops has been observed in electron photomicrographs.

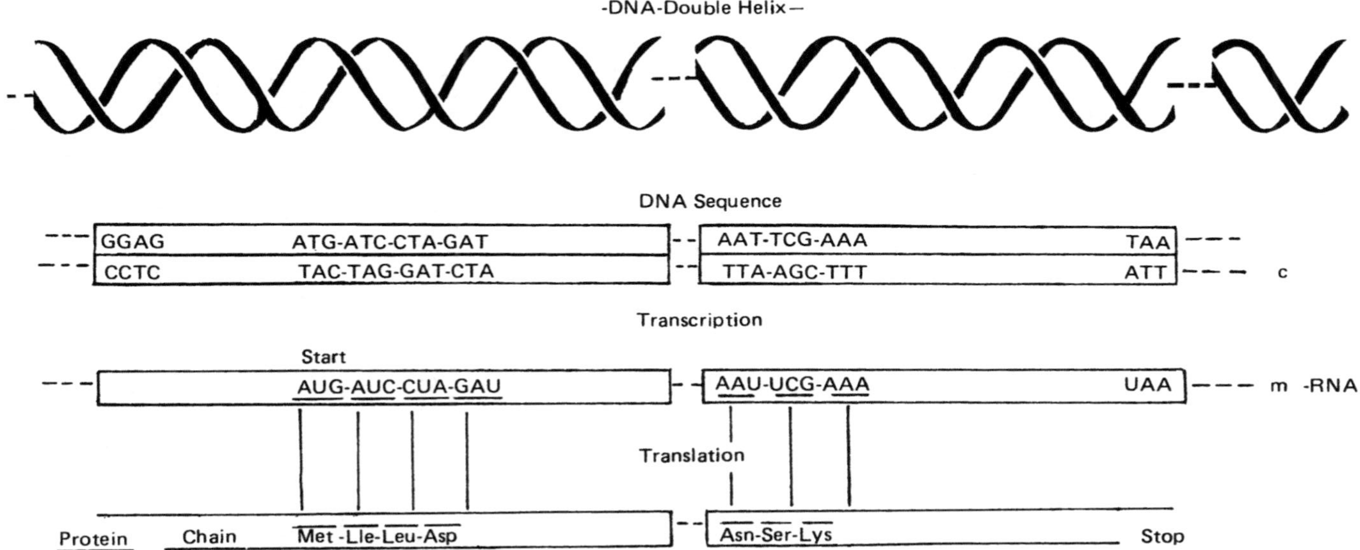

FIGURE 10 Transmission of genetic information in a bacterial cell. The transmission of the genetic information proceeds from the ds-DNA via base-pairing to the mRNA by "transcription," and is followed by the coding and formation of the polypeptide sequence by "translation." The detailed listing of the bases allows a step-by-step analysis of the individual transformations.

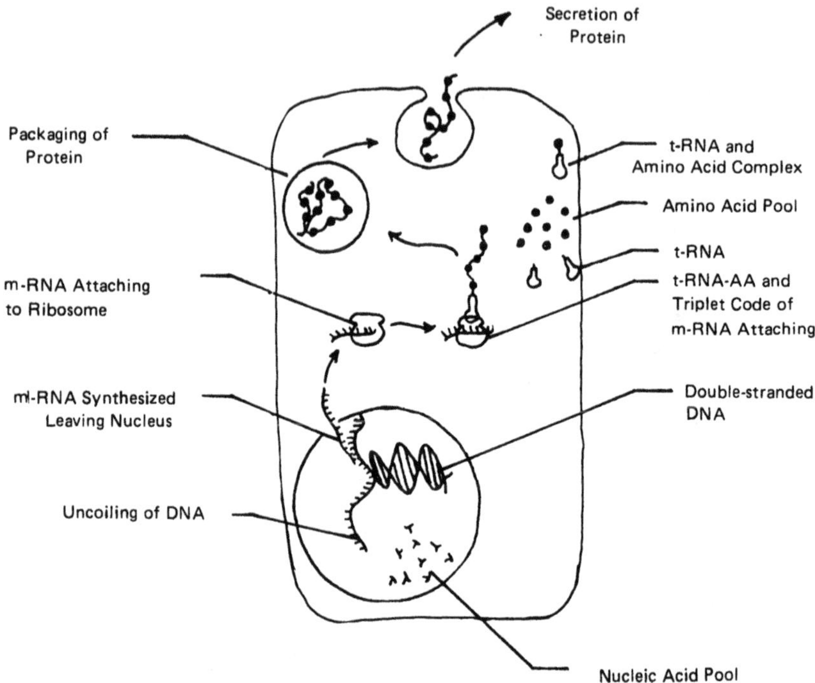

FIGURE 11 Pictorial illustration of protein synthesis in a eukaryotic cell. See text for description of sequences.

REFERENCES

1. Stent, GS. (1971). *Molecular Genetics*. W. H. Freeman & Co., San Francisco.

2a. Crick, FHC. *Prog. Nucl. Acid Res.* 1: 163 (1963); *Proc. R. Soc. B* 167:331 (1967).

2b. Woese, CR. (1967). *The Genetic Code, The Molecular Basis for Genetic Expression*. Harper and Row, New York.

3. Pelczar, MJ, Jr, Reid, RG, Chan, ECS. (1977). *Microbiology*, 4th Ed, McGraw-Hill, New York (1977); also Smyth, DG et al. *J. Biol. Chem.* 238 (1): 227–234 (1963).

4. Adams, RLP et al. (1981). *The Biochemistry of the Nucleic Acids*, 9th Ed. Chapman and Hall, New York.

5. Gassen, HG. The bacterial ribosome: a programmable enzyme (in German). *Angewandte Chemie* 94 (1): 15–27 (1982). in

6. Lake, JA. The ribosome. *Sci. Am.* 245 (2): 84–97 (1981).

7. Hopwood, DA. The genetic programming of industrial microorganisms. *Sci. Am.* 245 (3): 90–102 (1981).

8. Chien, Y-H, Thompson, EB. Genomic organization rat prolactin and growth hormone genes. *Proc. Natl. Acad. Sci.* 77 (8): 4583–4587 (1980).

9. Maurer, RA, Gubbins, EJ, Erwin, CR, Donelson, J. Comparison of potential nuclear precursors for prolactin and growth hormone messenger RNA. *J. Biolog. Chem.* 255 (6): 2243–2246 (1980).
10. Dickerson, RE, Geis, I. (1969). *The Structure and Action of Proteins.* Harper and Row, New York.

5
Enzymes

Enzymes have been called the "agents of life." Their function is to serve as catalysts for all biological processes. They are complex proteins and are extremely proficient in promoting the course of reactions. As catalysts, they accelerate chemical reactions, usually at temperatures considerably lower than those normally required for the reaction to proceed. While normal catalytic action implies participation of a factor in a reaction *without* being consumed, there are some instances where an enzyme apparently is used up in the process.

In biochemical terms the substance which is acted upon by an enzyme is called a substrate. Enzymes fulfill their function when present in very small amounts in the cell. Thus, even when a rather abundant enzyme is involved, it is likely to amount to less than 1% of the total organic matter of the cell. Usually, the amount is much more likely to be in the 0.1% range. Structurally, enzymes being proteins, are composed of some 200 to more than 1,000 amino acids. These acids are covalently linked in a pattern regulated by a cell's genetic code. The molecular weights will vary from 10,000 to 1,000,000. In essence proteins are long-chain polymers; coiled and folded in a specific structure to facilitate their unique function.

The protein chain of the enzyme "ribonuclease" was shown in Chapter 4, Figure 2; disulfide bridges (cysteines) between sections of the linear polymer are indicated by –S–S– bonds.

The action of enzymes as expressed by the Michaelis-Menten concept involves two steps: (a) formation of an enzyme-substrate complex, and (b) the actual chemical reaction which transforms the substrate held in the complex to the reaction product (Fig. 1).

Enzymes, for the most part are highly specific in their action. This specificity

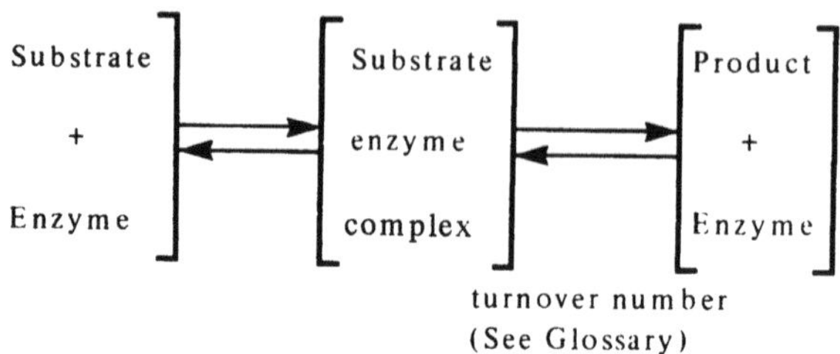

FIGURE 1 Enzyme–substrate complex.

is due in part to amino acid sequences, but to a much greater extent to the manifold three-dimensional configurations which are possible. The spatial positions of side chains, brought into proximity by chain contortion create so-called *active sites*, relatively small surfaces located in structural clefts, which then confer the observed specific activity (same concept as used with inorganic catalysts).

The tertiary structures of enzymes are quite complicated. Artists' illustrations of enzyme structures are presented by Dickerson (1) and Phillips (2), as well as in a number of texts on molecular biology. Such diagrams essentially convey the existence of a myriad of side-chain elements with indication of active-site creation. The situation is somewhat analogous to the concept of energy wells.

NUCLEASES

Many of the enzymes involved in recombinant DNA technology are nucleases, that is, they are enzymes that degrade nucleic acid. The prominent characteristics that define a nuclease are:

1. Substrate specificity: that is, action on DNA or RNA or both of the nucleotide chains.
2. Mode of attack: polynucleotide attack at points within the chain, that is, *endolytical* cleavage, or a stepwise attack at the chain end, that is exolytical cutting. Intrachain cutting by *endonucleases* produces oligonucleotides and such fragmentation has a pronounced effect on physical properties, as for instance in the viscosity of a DNA solution. The *exonucleases* nibble at the chain ends, producing mononucleotides and so have much less effect on the remaining chain properties.
3. Mode of phosphodiester bond cleaving: Because of the unique 5′ and 3′ orientational structure, there are two possible cleavage reactions.
   a. Cutting the bond between the 5′-OH group and the phosphate group by hydrolysis yields a 3′-phosphoryl end group.

b. Hydrolysis of the bond between phosphate group and the 3′-OH group produces a 5′-phosphoryl end group.

SITE-SPECIFIC RESTRICTION ENDONUCLEASES

Developments in DNA technology are totally dependent upon the findings that enzymes cut the DNA at specific sites. There are now two types of such enzymes. Thus, type I will cut one strand of DNA at a random site, that is its recognition sequence is asymmetric. For instance, restriction endonuclease EcoRI uses the recognition sequence:

G↓AA*TTC

where A* is *N*-6-ethyl adenosine.

The Type II enzyme recognizes a *particular target sequence* in a duplex DNA molecule and breaks the polynucleotide chain within that sequence or at a constant distance from the sequence, resulting in DNA fragments of defined length and sequence. The DNA sequence recognized by Type II enzymes is a palindrome, so called because it is symmetrical, and is usually 4 or 6 base pairs in length. The cleavage products may be blunt-ended (no single-stranded regions generated) or they may be staggered cuts. In the latter class the overlapping single-stranded ends may be easily rejoined with other DNA fragments with the same overlapping ends. If two DNA fragments from different sources are joined together after cleavage by the same enzyme, the resulting DNA is called "recombinant."

ENZYMES USED IN RECOMBINANT DNA TECHNOLOGY

Enzyme Nomenclature

Site-specific restriction endonucleases are designated such that the genus name of the host organism gives the first letter, followed by the first two letters of the species identification. The strain designation, where pertinent, follows the three letters. The letter R may precede the enzyme abbreviation, indicating that all enzymes are endonucleases, but often R is omitted. If a particular host strain has several different restrictions, they are indicated by Roman numerals. Examples are:

*E*scherichia *co*li strain RY13: *Eco*RI
*H*aemophilus *inf*luenza strain R: *Hinf*I
*H*aemophilus *in*fluenzae strain R*d*: *Hind*III (third described)

Some specific enzymes are:

SI nuclease: Marked preference for degrading (cleaving) single-stranded polynucleotides: DNA and RNA.

Alkaline phosphatase: Contained in *Escherichia coli*; it removes 5′-phosphate groups from RNA and DNA.

Lambda (λ) exonucleases: Remove nucleotides from 5′ ends of double-stranded DNA (ds-DNA); do not attack at nicks or gaps.

DNA ligase: Present in *E. coli* and is also coded for by the phage T4; it catalyzes the formation of a covalent phosphodiester bond from a 5′-phosphoryl group and an adjacent 3′-hydroxyl group. It repairs signle-stranded nicks in duplex DNA and joins DNA fragments with cohesive or blunt ends. These enzymes require either nicotinamide adenine dinucleotide (NAD) or adenosine triphosphate (ATP). DNA ligase is essential for the recombination of DNA in vitro.

DNA polymerase *I*: From *E. coli*, also called "DNA-dependent DNA polymerase"; converts single-stranded DNA to double-stranded form.

Terminal nucleotidyl transferase: Obtained from calf thymus; will add oligodeoxynucleotide tails, for example oligo (dT), to 3′ ends of single-stranded or double-stranded DNA.

Reverse transcriptase: Converts poly A^{+}RNA to ss-c DNA and on to ds-DNA under certain experimental conditions (see polymerase I), synthesizes cDNA template, that is, ss-DNA complementary to mRNA, that is, mRNA ⟶ cDNA.

Site-specific restriction endonucleases (type II): Cleavage of DNA at specific sites into specific fragments having single-stranded, complementary overhanging ends or blunt ends. Almost 400 restriction endonucleases have been discovered, with a minimum of 91 different specificities (1) (see also Table 1).

Host-Controlled Restriction and Modification: Significance of Restriction

The process of restriction and modification was first observed for phage infection of bacterial hosts. For example when the bacteriophage λ was grown on *E. coli* C, it reinfected *E. coli* C at high titer but was much less efficient in infecting *E. coli* K12. If the phage that did grow on *E. coli* K12 were used to infect the two strains, it reinfected *E. coli* K12 at high titer but was much less efficient in infecting *E. coli* C. It was discovered that the basis for this apparent anomaly was that *E. coli* K12 contained an enzyme that degraded DNA from phage propagated on *E. coli* C. Those few phage that escaped degradation had their DNA modified (by methylation of adenine or cytosine residues) by the host

TABLE 1 Cleaving DNA with Restriction Endonucleases

Restriction enzyme	Sequence recognized	Fragment end structure after cleavage	Mean fragment length (bp)
*Hin*fI	GANTC	$_{p}$ANTCNN GNN	256
*Hpa*II	CCGG	$_{p}$CGGNN CNN	256
*Sau*I	GATC	$_{p}$GATCNN NN	256
*Taq*I	TCGA	$_{p}$CGANN TNN	256
*Eco*RII	CC$^{A}_{T}$GG	$_{p}$CCXGGNN NN	512
*Ava*II	GG$^{A}_{T}$CC	$_{p}$GXCCNN GNN	512
*Ava*I	CYCGRG	$_{p}$YCGRGNN CNN	1024
*Bam*HI	GGATCC	$_{p}$GATCCNN GNN	4096
*Bgl*II	AGATCT	$_{p}$GATCTNN ANN	4096
*Eco*RI	GAATTC	$_{p}$AATTCNN GNN	4096
*Hind*III	AAGCTT	$_{p}$AGCTTNN ANN	4096
*Sal*I	GTCGAC	$_{p}$TCGACNN GNN	4096
*Xba*I	TCTAGA	$_{p}$CTAGANN TNN	4096
*Xho*I.*Blu*I	CTCGAG	$_{p}$TCGAGNN CNN	4096
*Xma*I	CCCGGG	$_{p}$CCGGGNN CNN	4096

The enzymes listed cleave DNA to produce symmetric, staggered cuts, leaving the 5′ ends extended. Enzymes used are T4 polynucleotide kinase or DNA polymerase. Subscript *p* preceding fragment structure indicates a phosphate at the 5′ end.
Source: From Ref. 4.

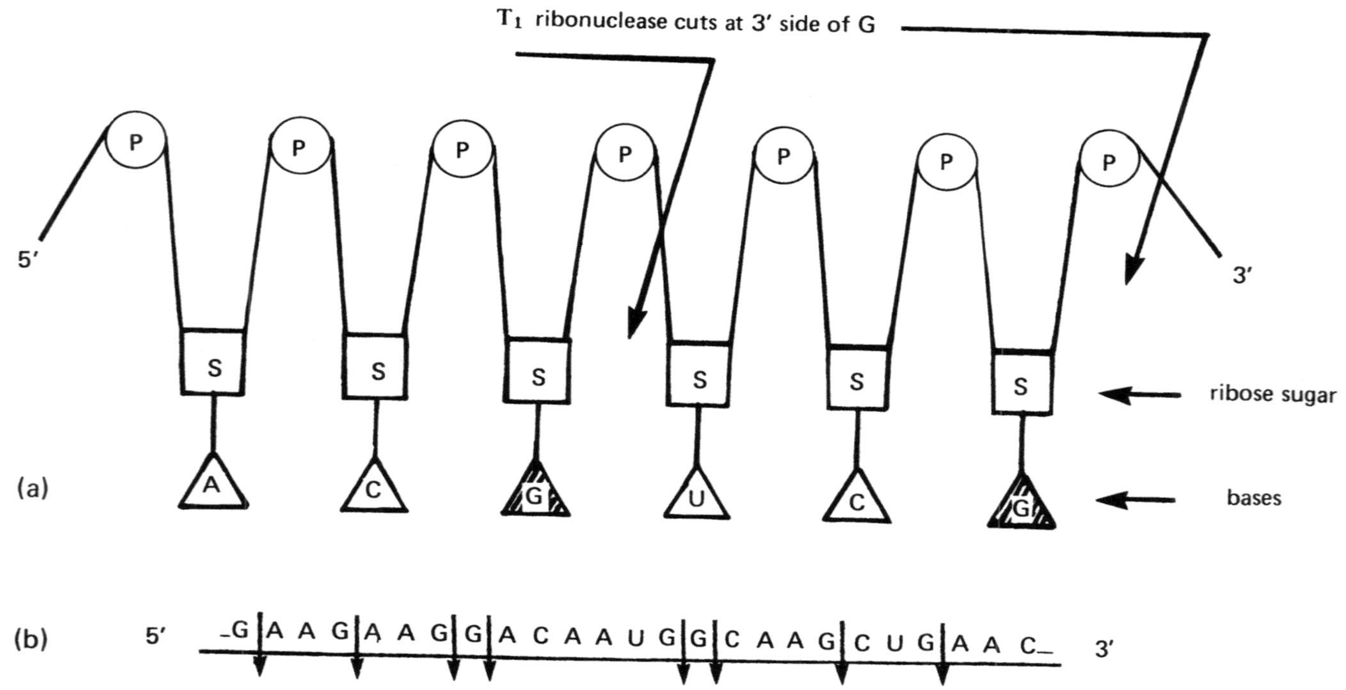

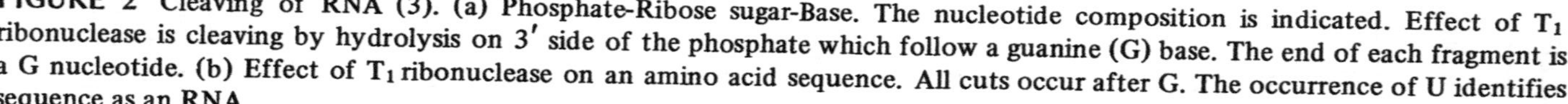

FIGURE 2 Cleaving of RNA (3). (a) Phosphate-Ribose sugar-Base. The nucleotide composition is indicated. Effect of T_1 ribonuclease is cleaving by hydrolysis on 3′ side of the phosphate which follow a guanine (G) base. The end of each fragment is a G nucleotide. (b) Effect of T_1 ribonuclease on an amino acid sequence. All cuts occur after G. The occurrence of U identifies sequence as an RNA.

E coli K12, making the phage efficient at infecting *E. coli* K12. When the modified phage was used to infect *E coli* C, it was degraded by endonucleases which recognized the methylated DNA as foreign. The endonucleases responsible for these degradations are called "restriction endonucleases" or "restriction enzymes," and they protect the host from foreign DNA.

CLEAVING OF RNA

Enzyme cleaving of RNAs is entirely analogous to that of DNA cutting. Of course, the enzymes must recognize the uracil (U) base instead of the thymine (T) molecule. The action of the T_1 ribonuclease (enzyme) on a RNA nucleotide sequence is diagrammed in Figure 1.

REFERENCES

1. Dickerson, RE. The structure and history of an ancient protein. *Sci. Am.* 226 (4): 58–72 (1972).
2. Phillips, DC. The three-dimensional structure of an enzyme molecule. *Sci. Am.* 214 (5): 78–90 (1966).
3. Woese, CR. Archaebacteria. *Sci. Am.* 244 (6): 98–122 (1981).
4. Roberts, RJ. (1980). Directory of restriction enzymes. In *Methods in Enzymology, 65*. Grossman, L, Moldave, K (Eds.), Academic Press, New York, Part I, pp. 1–15.
5. Maxam, AM, Gilbert, W. Sequencing end-labeled DNA with base-specific chemical cleavages. In *Methods in Enzymology, 65*. Grossman, L, Moldave, K (Eds.), Academic Press, New York, Part I, pp. 499–560.

6
Plasmids, Viruses, and Microbial Hosts—Vectors

The relationship of plasmids, viruses, and bacterial hosts is an intimate one and frequently constitutes a cyclic event. For instance, one cultures *Escherichia coli* bacilli to obtain the plasmids contained in the cell, then removes these plasmids and uses them as a vector to convert them into a recombinant structure, and finally reintroduces the new plasmid back into *E. coli* bacteria as the microbial host for cloning. When a viral phage is used, the procedure is slightly different, in that the viral DNA or RNA is isolated, recombinantly modified, and then inserted into a bacterial host for cloning.

PLASMIDS

A concise definition of plasmids, as presented for instance by Bernard and Helinski (1), states that they are "extrachromosomal, self-replicating and stably-inherited nucleic acid molecules always existing as double-stranded DNA." Plasmids are predominantly circular elements, perhaps better said looped structures. However, some linear plasmids were reported in 1979 (2) and 1981 (3). The most recent account is that of Hirochika and Sakaguchi (4), who sequenced linear plasmids isolated from *Streptomyces* spp.

Plasmid Culturing

Bacterial stocks are usually maintained at -70°C. When a culture is to be prepared for laboratory growth, a number of flasks, 500 ml to 1 liter size, are filled partially with the growth medium or broth. The flasks are then inoculated with the stock culture. A frequently used bacterium is *E. coli*. The inoculated flasks are maintained at culturing temperature, 37°C, in an incubator, perhaps overnight. To prevent contamination the mouth of the flask is covered with a sponge-type stopper. The first cell division occurs in about 60 min, and the multiplication proceeds exponentially:

$N = 2^{t/T} N_o$

N = Size of bacterial population after t = elapsed time

T = Generation period (20-50 min.)

N_o = Number of bacteria at start

Expressed in log terms:

$$\log_{10} N = \frac{0.301}{T} t + \log_{10} N_o$$

When a sufficient amount of bacteria has been grown, the bacteria are separated from the broth by centrifuging at ~6,000 x g, resuspended, and washed repeatedly until the nutrients have been removed to the desired degree. The preparation is then ready to undergo cell rupturing (lysing) and further plasmid purification.

Chloramphenicol Enrichment

A rather neat procedure is the use of chloramphenicol (Cm, an antibiotic) addition in plasmid culturing. As plasmid-bearing bacteria, such as *E. coli* are being cultured, the addition of Cm results in the inhibition of cell growth and protein formation, while the plasmid DNA continues to proliferate. With the creation of as many as 1,000 to 3,000 copies of the plasmid per cell, the resulting enrichment is substantial. This procedure is only effective for plasmids, such as the *Col*El plasmid, whose replication does not require protein synthesis. These plasmids are relatively rare. Note, *Bacillus subtilis* cloning of plasmids does not respond to Cm enrichment in additional overnight incubation.

Plasmid Sources

Among common bacterial strains which can be harvested for plasmids are varieties of *E. coli*. Table 1 lists plasmids of different molecular weights (MW) which can be obtained from a MW as low as 3.1×10^6 to a relatively large size of 13.8×10^6. *E. coli* contains plasmids of much larger size, such as the conjugal plasmid F, which has a molecular weight of 62×10^6. Interestingly, strain SPA-O does not produce plasmids.

TABLE 1 Bacterial Strains to Obtain Plasmids

E. coli strain	Plasmid	$10^{-6} \times M^r$
581 (*thy*$^-$, *str*^{R})	ColE1	4.5
OS410 (*str*^{R}, *thi*$^-$)	Δ 362	3.1
OS410 (*str*^{R}, *thi*$^-$)	ColE1::Tn7	13.8
SPA-0	none	–
C6H27 (*trp*^{P}, *Bl*$^-$, *str*^{S})	pML21	7.9
HB101	pXlr101	10.4
HB101	pXl108	12.7
HB101	pXl212	11.5

Strains 581, OS410 and C6H27 were a kind of gift of D. Sheratt. The molecular weights of their plasmids were determined to be approximately 1.07 times their published values (D. Sheratt, personal communication.
Source: From Ref. 5.

Preferred properties of plasmids for use as vectors are:

Small size (low MW).
Readily identifiable markers (e.g., resistance to drugs or antibacterial agents, preferably two such markers).
Stability in a bacterial host.
Replication and amplification capabilities in the host.
Single-restriction endonuclease cutting sites within resistance genes (antibiotic).
Non-conjugativity (i.e., no transfer of DNA between individual cells).

Basic Properties

Plasmids are present in many prokaryotes (bacteria) and, carry a great variety of genes. Properties owed to such genes are commonly referred to as phenotypic traits. Among such expressed functions are the well-known antibiotic resistance genes such as tetracyline resistance. Tc^r or TET^R, as well as resistance to ampicillin, kanamycin, streptomycin, or multiple drugs. The antibiotic resistance property is frequently used to screen the recombinant entities prior to cloning. In Chapter 10, dealing with applications, a "super bug" is described which takes advantage of microbial abilities to metabolize hydrocarbons. Also, genes responsible for the induction of plant tumors do exist, and this matter is expanded in Chapter 9. A few examples of plasmid properties are summarized in Table 2.

TABLE 2 Properties of Some Common Plasmids

Source: *E. coli*	Size kilobases	Marker genes	Single restriction enzyme sites
pBR322	4.3	Amp^r Tet^r	*Eco*RI, etc. (see Chap. 7, Fig. 4).
pBR313		Amp^r Tet^r	*Eco*RI, *Bam*HI, *Sal*I *HindIII*, *Sma*I, *Hpa*I
pACYC 177	3.7	Amp^r Kan^r	*Bam*HI, *Sma*I, *Hind*II *Xho*I, *Hind*III, *Pst*I
pMB9	5.2	Tet^r ColEI imm.	*Bam*Hi, *Hind*III, *Sal*I *Eco*RI
pUR	2.7	Amp^r Lac Op. Z gene	*Pst* I, *Pvu* I *Eco* RI
pSC101	5.8	Tet^r	*Eco*RI (considered by some investigators as a natural plasmid)
pUC8 and pUC9	2.8	Amp^r lac α-segment	*Eco*RI, *Sma*I, *Xma*I, *Bam*HI, *Sal*I, *Acc*I, *Hind*III

Two major types of plasmids, conjugative and nonconjugative, are described:

Conjugative plasmids carry tra genes, which promote *bacterial conjugation*, that is, the ability of transferring pieces of DNA from the donor, which contains the conjugal plasmid, to an appropriate recipient bacterium.

Nonconjugative plasmids, for instance are *Col*EI and RSF 1030. Conjugative types are RI, R6, and Ent P 307. which interestingly all possess multiple drug resistance.

Two general types of plasmids are also differentiated for their respective properties: (see Table 3).

1. Plasmids capable of high copy number (10–40 per cell), but are restricted to a very small number of bacterial species, for instance, pBR322 in *E. coli.*
2. Derivatives of plasmids that can be maintained in a large number of bacterial species (broad host range), but occur in relatively low copy number (1-5 per cell), for example, R6.

High copy number is more useful for cloning protein or peptide genes in a suitable host. Low copy number is better suited to modify a "commercial microbe" (cloning of useful genes).

One of the preferred artificial plasmids is pBR322. It is currently the most versatile hybrid plasmid for the following reasons:

1. It is small (4.3 kb) and thus has fewer restriction sites; also the cloning yield of fragment is maximized.
2. It is derived from the *E. coli* plasmid, *Col*El and can be amplified thus increasing the DNA yield.
3. It has 5 unique single restriction sites which can be used for insertion of ds-cDNA: *Hind*III, *Bam*HI, *Sal*I, *Pst*I, *Eco*RI.
4. It contains two selective markers, ampicillin and tetracycline resistance, they are not transposable elements (do not move within the genome).
5. It has been sequenced in its entirety.

A pair of plasmids, pUC8 and pUC9, has been described recently by Vieira and Messing (6). The pUC plasmids, an M13mp7-derived system for insertion mutagenesis and sequencing with synthetic universal primers, are also extremely useful as cloning vectors. They are small (2.8 kb), are amplifiable since they are derived from pBR322, and they contain 7 single restriction sites (see Table 2). They have two additional features that make them particularly useful. (a) These plasmids contain a small region (~400 bp) that contains the *lac* promoter and part of the amino terminal portion of β-galactosidase (*a* fragment). In appropriate *E. coli* strains that produce only the carboxyl terminal portion of this enzyme, the two fragments can assemble to form a functional β-galactosidase enzyme (this enzyme is normally a tetramer). Thus the presence of the pUC plasmids will convert some Lac^- *E. coli* strains to Lac^+. The single cloning sites listed in Table 2 are all within the region coding for the *a* fragment. Insertion

TABLE 3 Characteristics of General Classes of Plasmids

Characteristic	Small plasmid	Large plasmid
Autonomous replication	Yes	Yes
Conjugative (Tra)	No	Usually yes
Copy number	10–40 per cell	1–5 per cell
Amplifiable	Sometimes yes	No

of recombinant DNA into any of these sites will convert the phenotype of plasmid-containing strains to Lac$^-$. Thus, when strains of *E. coli* are grown on the appropriate media, colonies containing plasmids with inserts can be directly visualized by their Lac$^-$ phenotype. (b) The DNA inserted into these pUC plasmids can be easily sequenced by the chain-terminator nucleotide sequencing method (7) in the presence of single-stranded M13 primer (8).

Isolation of Plasmids

The bacterial cell (see Chap. 2, Fig. 1) contains plasmids suspended in the cytoplasma as well as chromosomal DNA. To remove the plasmids (circular DNA)* the cell membrane has to be opened. This can be done mechanically by shearing or ultrasonics, osmotically by bursting, or by lysing, that is dissolution. In any case, a mixture of suspended particles (cell fragments, chromosomal DNA of varying molecular weights, and plasmid DNA) will ensue. Mechanical shearing and ultrasonics will also shear the DNA, and these procedures, therefore, are not used when intact plasmid or high-molecular-weight chromosomal DNA is desired.

The most often used, and essentially, the conventional procedure, is lysing. The aim is to carry lysis to a point where the plasmids are liberated together with a limited amount of high-molecular-weight DNA. Then, high-speed centrifugation will remove cell debris and the high molecular weight DNA. This will give a "cleared lysate." The procedure often calls for a combination of "skill, luck, and patience."

Plasmid Purification: Buoyant Density Gradient Method

While there are several methods available to isolate pure plasmid DNA from cleared lysate, the most frequently used method is gradient centrifugation. This procedure is an isopycnic centrifugation. The cleared lysate is dispersed in a CsCl solution with addition of ethidium bromide (EtBr is mutagenic) (see Appendix 1 for EtBr structure).

EtBr intercalates between the DNA base pairs and causes the DNA to unwind. The density of the DNA-EtBr complex decreases as more EtBr is taken up. The plasmid DNA continues to take up EtBr until it is converted from a negative super coil to a positive super coil of the same density. Shearing will give a linear chromosomal DNA complex, which is then appreciably less dense than the covalent circular DNA of the plasmids.

The CsCl-EtBr-DNA solution is then transferred to a 40 ml polyallomar tube,

---

*Some chromosomal DNA of *E. coli* is also circular.

nearly filled to capacity, capped, and heat-sealed. The sealed tube is spun for 48 to 60 h at 40 K rpm. This establishes a gradient and separation is achieved. The lighter DNA-EtBr complex will be located in a band which lies above the heavier plasmid DNA. The technician then inserts a pipette through the opening at the top of the plasmid layer, or penetrates the tube, laterally, and aspirates the layer (Fig. 1). Also useful is a perforation in the bottom of the tube and use of an automated fraction collector. Subsequently, the EtBr must be removed by repeated solvent extraction. All operations are carried out with extreme care; wearing protective gloves, allowing *no* spillage to avoid EtBr contamination of the skin. Quantities of plasmids recovered are in the range of 100 μg to 2 mg per liter of culture. Figure 2 shows the automatic collection scheme for fractions of the cell product in gradient separation. The location of different density bands is shown in Figure 3.

This method of plasmid isolation is time consuming and somewhat inefficient. Admittedly, the isolation process in any case is not an easy procedure, but it appears that an alternative method, could represent improvements.

The problem of plasmid isolation is of considerable importance, especially the scaling-up of possible procedures.

Therefore, besides the well-established gradient separation, other processes are receiving attention, in particular procedures which would circumvent the potentially troublesome use of the mutagenic ethidium bromide. Presently, two processes have been recommended, both of them based on the selective adsorption of DNA molecules on fixed bed adsorbants. Both methods are the same in the physical sense, and so far differ only in the type of adsorbent used. Thus, work with hydroxyapatite ion exchange and differential elution has been reported by Colman, and the use of a relatively new Sephacryl medium (exclusion chromatography) is recommended by Pharmacia Fine Chemicals AB.

CHROMATOGRAPHIC PLASMID ISOLATION

Growth of Bacteria (Colman Study)

The bacterial strains used in Colman's study (5) are listed in Table 1. Colonies of the strains, except HB101, were "picked" from a freshly "streaked" agar plate into 10 ml of nutrient broth and grown over night at 37°C. Each culture was then used to inoculate 1 liter of either *L* broth or *M-9* broth, to which were added 1.5% of casamino acids and 2% glucose. These cultures were grown at 37°C to an "absorbance" of 0.6 at 650 mm (optical reading corresponding to 3×10^8 cells/ml). Solid chloramphenicol was added to a final concentration of 100 μg/ml. Incubation was continued for another 20 h.

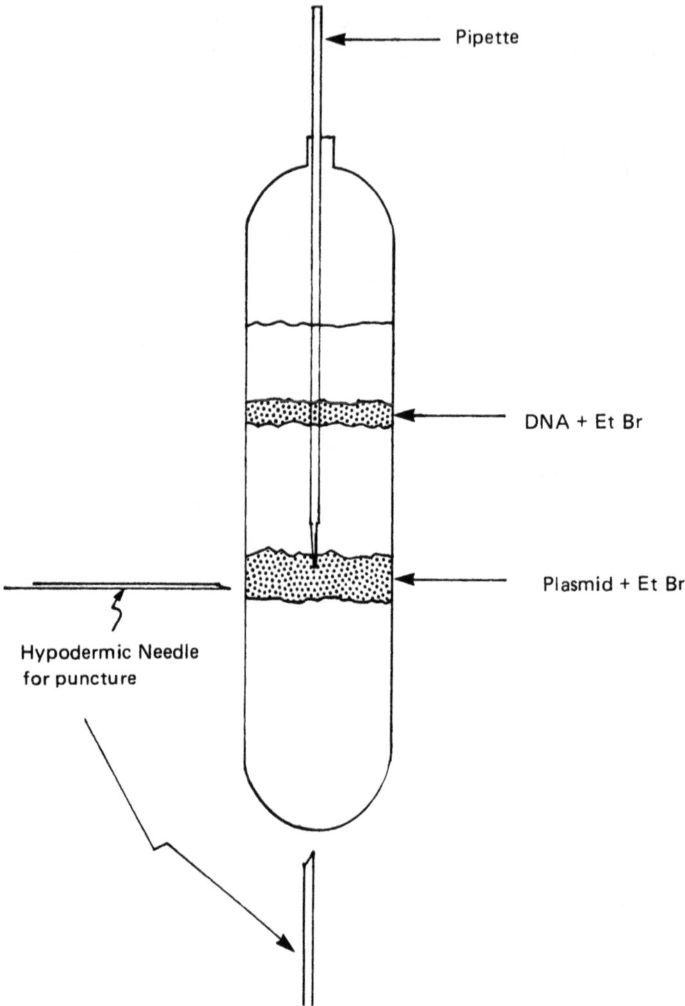

FIGURE 1 Diagram of separation container. Conventionally, 40 ml polyallomar tube is used Diagram shows "density bands" after centrifugation, and schemes used to collect plasmids.

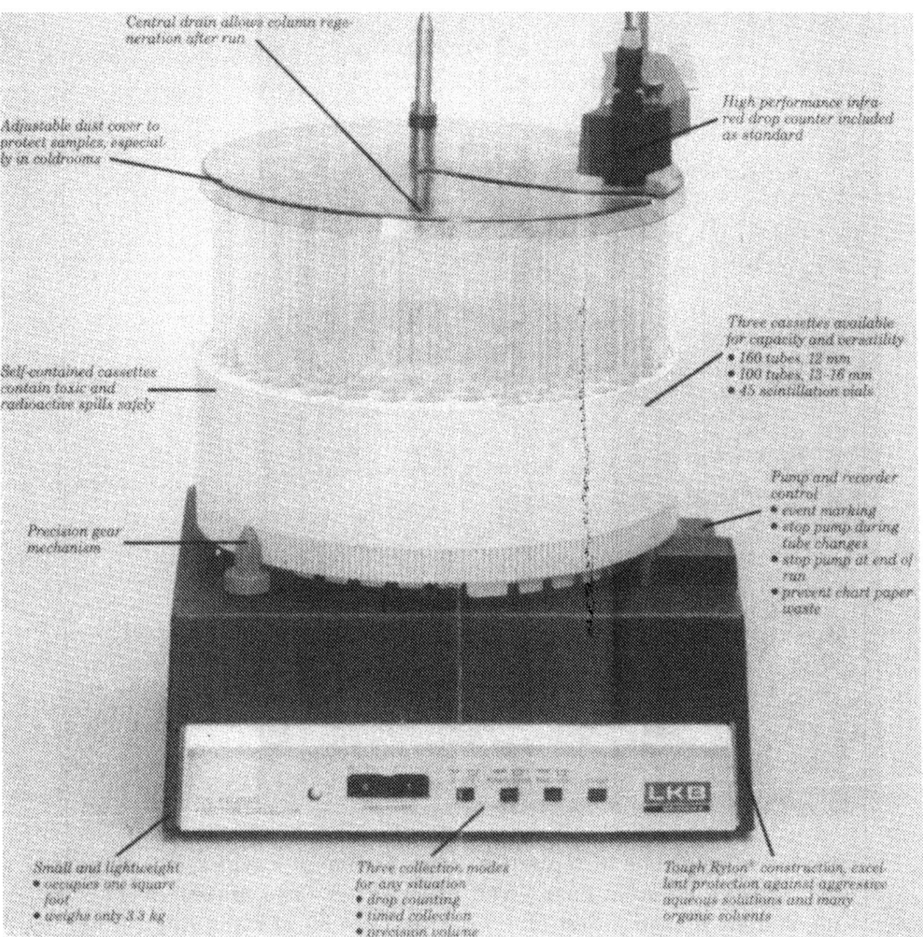

FIGURE 2 Automatic collection scheme in density gradient separation. Differences in density of DNA molecules results in stratification; after centrifugation is completed, the bottom of the tube is punctured and "automatic" collection is used. (Courtesy Pharmacia-LBK)

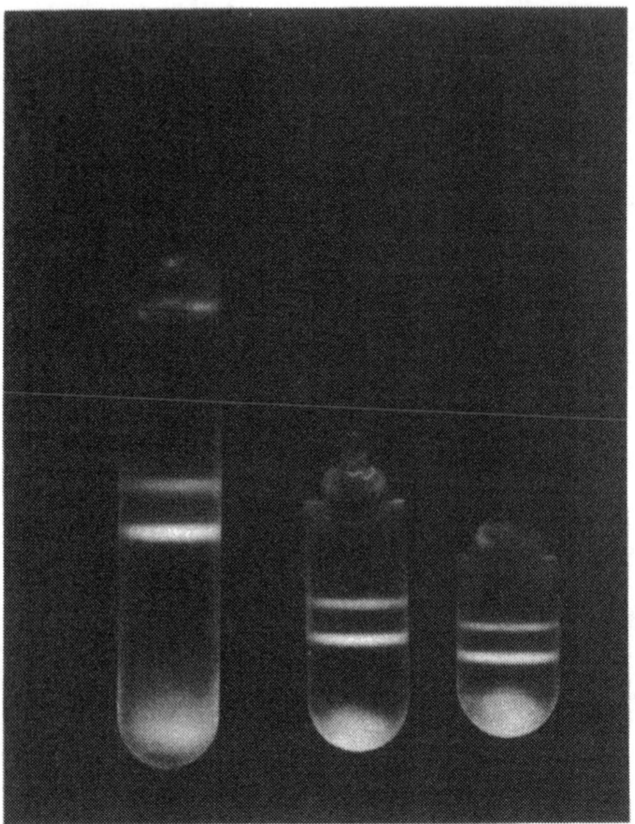

FIGURE 3 Photograph of separation tubes. Picture of a typical plasmid preparation using an ultracentrifuge. The figure shows three tube sizes (13.5 ml, 6.3 ml, and 4.2 ml Quick Seal tubes) run on Type 80 Ti Fixed Angle Rotor in an L8–80 M. The lower band is plasmid DNA. The upper band is chromosomal DNA. These were run in a CsCl ethidium bromide gradient. Average density was 1.55 g/ml. (Courtesy Beckman Instruments, Inc.)

Preparation of Cleared Lysate

As reported by Clewell and Helsinki (9): cells were harvested from 30 ml of culture and resuspended in 1 ml of cold 25% sucrose and 0.05 *M* Tris, pH 8. Lysozyme (0.2 ml of 1 5 mg/ml solution in 0.25 *M* Tris, pH 8.0) was added. The suspension was maintained for 5 min at 0°C, then 0.4 ml of EDTA (0.25 M, pH 8.0) was added. The suspension was kept at 0°C with occasional swirling for another 5 min, after which lysis was brought about by adding 1.6 ml of a detergent mixture consisting of: 1% Brij 58, 0.4% Na deoxycholate, 0.0625 *M* EDTA, and 0.05 Tris, pH 8.0. After 5-10 minutes the samples became relatively clear and viscous. They were then centrifuged at 2°C for 25 min at 48,000 x g. This procedure usually "pelleted" about 95% of ^{3}H-labeled DNA. The supernatent layer, which contained the protein complexed ColEl DNA was called the "cleared lysate."

Hydroxyapatite Chromatography

The basic concept of hydroxyapatite chromatography is that large linear DNA (chromosomal DNA and nicked plasmid DNA) are selectively retained. Hydroxyapatite (HA) (Biorad DNA grade) was resuspended in 8 *M* urea, 0.24 sodium phosphate buffer pH 6.8 using 1 g HA for every liter of bacterial culture. Slurry was poured into 1 3 cm diameter glass column and washed with the phosphate (P) buffer. The procedure and all subsequent steps were carried out under a slight positive air pressure of 23 mmHg (3000 Pa). Next 1.1 volumes of the cleared lysate were diluted with 8.9 volume of 9 *M* urea, 0.27 *M* phosphate buffer, 0.9 sodium dodecylsulfate (surface active agent), adjusted to pH 6.8. This mixture was loaded on to the HA column.

The column was then washed with 8 *M* fresh urea, 0.24 *M* phosphate buffer, pH 6.8, until the absorbance at 260 nm (optical reading) of the column effluent was approximately zero. Two column volumes of 0.01 *M* phosphate buffer at pH 6.8, were then washed through the column. The HA was then resuspended in a further 2 volumes of the low phosphate buffer before a final 2 vol of buffer was washed through.

Addition of 0.3 *M* buffer at pH 6.8 caused the plasmid DNA to elute. DNA recovery was monitored either spectrophometrically or chemically, using a diphenyl amine assay. Different batches of HA gave varying recoveries from as much as 5 mg plasmid DNA in a column containing 3 g of HA to as little as 0.6 mg.

Gel Electrophoresis

The procedure was done in 0.8% agarose gel (see Chap. 8).

All detectable protein and RNA contamination of plasmids was removed and the conformation (shape) of the plasmid DNA was unaffected. Less than 0.5%

chromosomal DNA was present in the purified preparation. This was accomplished by including a heat denaturation step on the cleared lysate prior to HA chromatography: at 100°C for 2 min and rapidly freezing to -70°C in solid CO_2 propanol mixture.

Sephacryl Gel Filtration

The Sephacryl bulletin (Pharmacia, Ref. 10) describes the plasmid and chromosomal DNA (chopped up DNA) separation as gel filtration. The principle of this separation process is illustrated in Figure 4. Sephacryls are cross-linked polymers of allyl dextran with *N,N′*-methylene bisacrylamide. The product can be made with a well-controlled pore size and is a rigid gel; head diameter is 40-105 μm. An example is as follows: pBR plasmids from *E. coli* 259 cells. After lysis and removal of cell debris (centifuging), the bulk of the chromosomal DNA was removed by denaturing at alkaline pH and precipitation. The supernatant was treated with *ribonuclease* A, followed by phenol extraction (to separate proteins). A 250 μl sample containing ~ 100 μg plasmid material was applied to a packed column of "Sephacryl S-1000 Super Fine" (bed height 30 cm, diamter ~ 2 cm). The eluent solution was Tris-phosphate pH 8 containing 1 *M* NaCl. Flow rate was 16 ml/h. Plasmid recovery was > 90%. Electrophoresis for purity was performed on 0.8% agarose gel. Figure 5 shows DNA concentration with time and gel banding.

VIRUSES

Viruses are the simplest biological entities. As such, however, they are not free-living structures, but lead a parasitic existence infecting host cells and using the metabolic functions of the cell to proliferate. They inhabit the cells of microorganisms, plants, and animals. A complete viral particle, called a virion consists of a basic block of genetic material, DNA or RNA, which is surrounded by a protective coat of protein called the capsid. Most bacterial viruses have only the protein coat, while viruses which infect eukaryotes are often covered by an outer lipid envelope. Taxonomically, the presence or absence of such an envelope separates the viruses into two categories: enveloped and nonenveloped (or naked). The lipid bilayer and associated proteins are important components which enable the virus entity to adhere to the host cell surface and to penetrate the cell wall. An electron photomicrograph and a structural diagram are shown in Chapter 2 Figure 4.

The nucleic acids are responsible for the infectious capabilities of the viruses. Virion sizes can vary from ~ 20 to 300 μm, so the magnifications needed in making electron micrographs range from 20,000- to 100,000-fold.

After the cells become infected with the viral nucleic acid, the complete viral

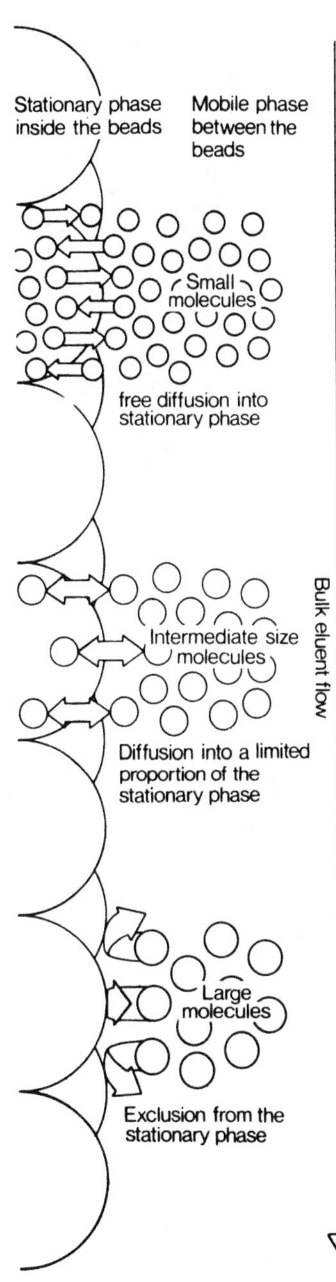

Gel filtration – Principle

Figure 1 illustrates schematically the principle of gel filtration. Molecules are eluted in order of decreasing molecular size.

Expression of results

The elution position of a substance can be characterized by its K_{av}

$$K_{av} = \frac{V_e - V_o}{V_t - V_o}$$

where V_e is the volume of elution of the peak maximum, V_o is the void volume and V_t is the total bed volume.

Some characteristics of gel filtration

- The capacity of a column is limited by the volume of the sample and not usually by its concentration.
- The separation occurs independently of the elution conditions, provided these do not alter the size or conformation of the molecules being separated. Elution conditions, therefore, can be chosen to suit the molecules being separated.
- Peak width increases with increasing flow rate i.e. optimum resolution is obtained at low flow rates.
- Resolution increases with increasing bed height.

FIGURE 4 Gel filtration. Description as shown. (Courtesy Pharmacia Fine Chemicals, Ref. 10.)

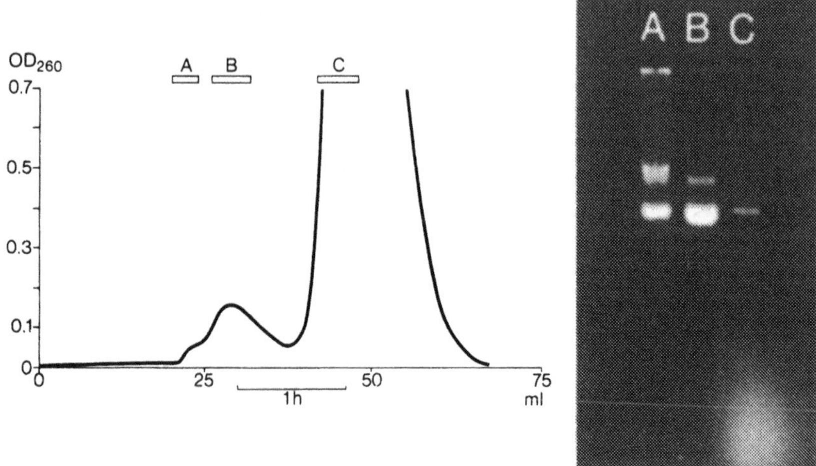

FIGURE 5 Purification of plasmids on sephacryl resin. Resin used was Sephacryl S-1000 Superfine. Electrophoresis results on 0.8% agarose gel. Lane A: residual amounts of chromosomal DNA, open plasmids and closed plasmids. Lane B: distinct band of closed plasmids and fainter band of open plasmids. Lane C: mainly RNA and nucleotides. (Courtesy Pharmacia Fine Chemicals, Ref. 10.)

particles are synthesized. When the virus infection occurs in bacteria, a so-called bacteriophage-bacteria system is formed. Conventionally, the term bacteriophage is abbreviated to "phage." The chemical composition of a bacterial virus suspension has been reported as 50 to 90% protein, 5 to 50% nucleic acid, and less than 1% of lipid, all on a dry weight basis. Those viruses with a lipoprotein sheath, however, may contain up to 25% lipid. Population densities in cultures are in the 10^9 to 10^{11} virus particles/ml range.

An informative summary of virus characteristics is presented in Table 4 and a diagrammatic comparison of sizes is shown in Figure 6.

Isolation

Bacteriophages and animal viruses are particularly useful in recombinant work. They represent the source of viral DNA. After the coat or envelope of the virus particle is destroyed (i.e., by lysing), the nucleotide components are separated in a manner similar to that used for plasmid isolation. The most commonly used procedure is the byouant density gradient method. If the phage mixture contains bacterial DNA, the density difference between it and the viral nucleotides permit a separation.

Cell and Virus

Ordinarily, the phage cycle ends in cell lysis. This happens at a late stage of virus reproduction, and the cell disruption is not ascribed to bursting caused by overcrowding by virus particles. It is thought that lysis of the cell wall actually occurs from both sides, that is destruction or erosion from within and also on the outside of the cell due to presence of viral particles from other lysed cells. The life-cycle behavior of animal viruses, from entering a cell, proliferation within the cell, and departure of new virus particles from the cell, is treated in minute detail by Simons et al. (12). Diagrammatic presentations illustrate all of the phenomena, which are quite complex in their sequences, including a pathway through the cellular ribosome to the endoplasmic reticulum and to the Golgi apparatus.

Bacterial Viruses

As previously stated, bacterial viruses are of prime interest in recombinant procedures. Two viruses which have been used consistently are bacteriophage lambda and phage M13.

Phage λ

Phage λ is a virus which infects *E. coli*, so it is a coliphage. It is particularly suitable for recombinant procedures. As reported by various investigators, the DNA of phage upon isolation from phage particles is a linear double-stranded molecule containing about 49 kilobase pairs. At each of the ends of the chromosome (gene sequence) there are 5′ projections of 12 nucleotides. These ends have complementary sequence which facilitate circularization. The sequences, thus, represent cohesive termini and when closure takes place it results in the formation of a *cos site* in a circular structure.

Plasmids containing a fragment of λ DNA with the *cos* site are called *cosmids*. They are suitable as gene-cloning vectors with a so-called in vitro packaging procedure. The map of the chromosome of phage λ is shown in Figure 7. Another area which is of interest to recombinant work is the "nonessential region," a region which is not essential for phage outgrowth. Thus it can readily be excised by means of appropriate enzymes and the gap is utilized by insertion of foreign DNA fragments. The notations "head" and "tail" mean that these regions code for the proteins in the head and the tail of the virus particle, respectively.

A truly comprehensive and very informative publication by Williams and Blattner (13) provides a concise explanation of how lambda can be used to clone DNA. Additionally, an extensive catalog of genetic configurations of λ vectors is presented, collected from materials available from various laboratories–the gene maps are like the one shown in Figure 7, but they present specific restriction

TABLE 4 Characteristics of Viruses

Morphological class	Nucleic acid[a]	Example		Size of capsid (A)	No. of capsomers	Size of virions of enveloped viruses	Special features
		Virus family	Virus				
Helical capsid							
Naked	DNA	Coliphage fd		50 × 8000			Single-stranded DNA
	RNA	Many plant viruses	Tobacoo mosaic	175 × 3000			
			Beet yellow	100 × 8000			
Enveloped	RNA	Myxoviruses	Influenza	90 diamter		900–1000	Fragmented RNA
		Paramyxo-viruses	Newcastle disease	180 diamter		1250–2500 and over	
		Rhabdoviruses	Vesicular			680 × 1750	Bullet shape
Icosahedral capsid							
Naked	DNA	Parvoviruses	Adenosatellite	200	12		Single-stranded cyclic DNA
			Coliphage ϕx 174	220	12		
		Papovaviruses	Polyoma	450	72		
			Papilloma	550	72		Cyclic DNA
		Adenoviruses		600–900	252		
		Tipula irridescent virus (insects)		1400	1472		

	RNA	Coliphage F2 and others		200–250			
		Picornaviruses	Polio	280	32		
		Many plant viruses	Turnip yellow	280	32		
		Reoviruses		700	92		Double-stranded RNA
		Wound tumor virus (plant)					
Enveloped	DNA	Herpesviruses	Herpes simplex	1000	162	1800–2000	
Capsids of binal symmetry (i.e., some components icosahedral, others helical)							
Naked	DNA	Large bacteriophages	T2, T4, T6	Modified icosahedral head 950 × 650; helical tail, 170 × 1150			
Complex virions	DNA	Poxviruses	Vaccinia		2500 × 3000		Brick shape
			Contagious postular dermatitis of sheep		1600 × 2600		

[a]DNA double-stranded, RNA single-stranded, unless specified in last column.
Source: From Ref. 11.

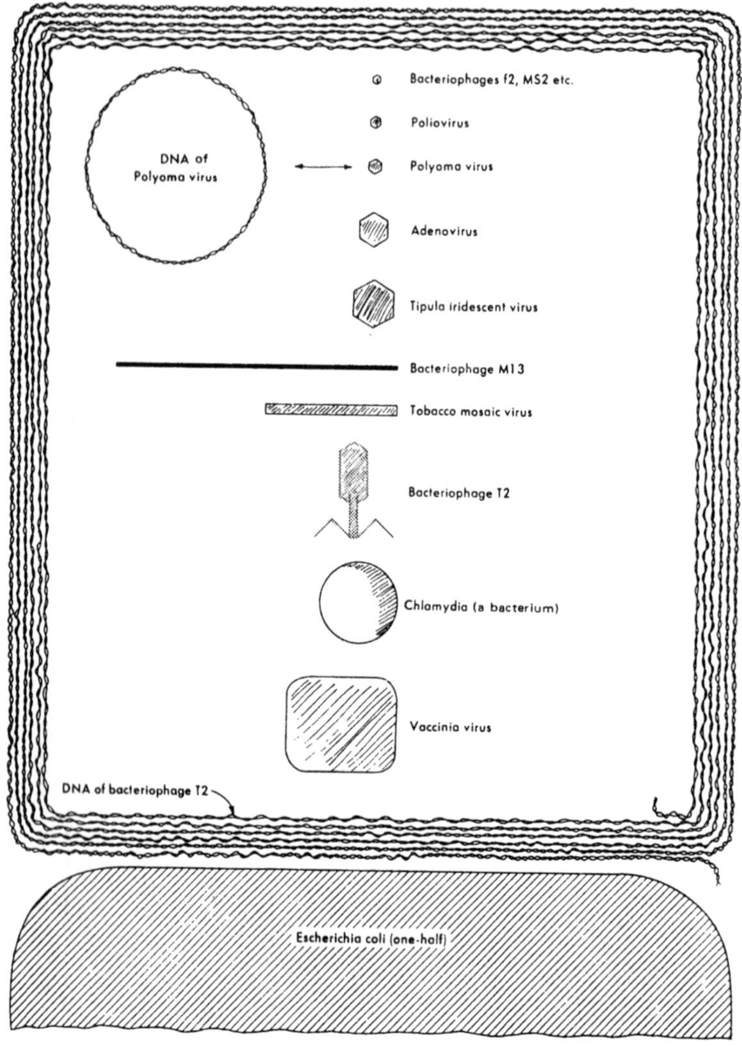

FIGURE 6 Comparative sizes of viruses. Comparative sizes of virions, their nucleic acids, and bacteria. The profiles, as well as the length of the DNA molecules, are reproduced on the same scale. (From Ref. 11.)

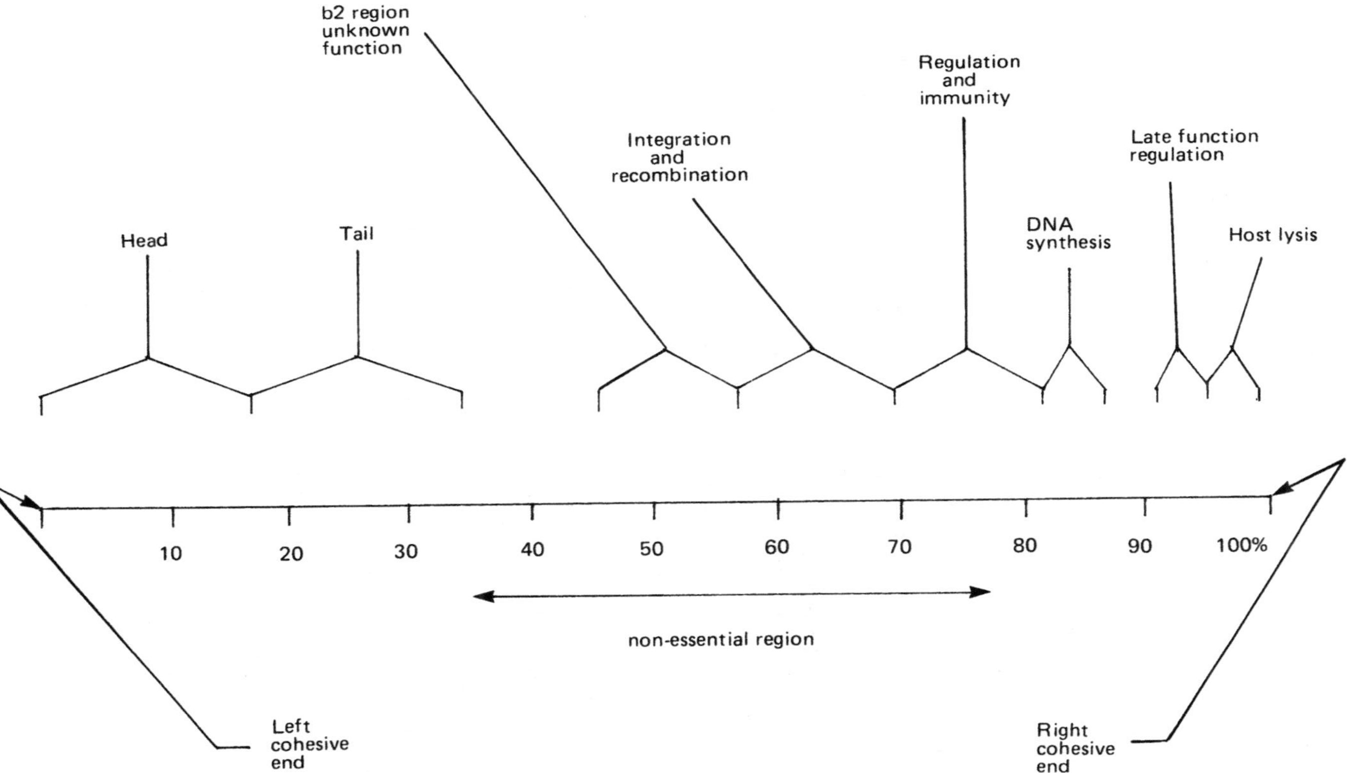

FIGURE 7 Map of phage lambda chromosome. Locations of functional genes.

sites in great detail. Furthermore, a wealth of data is given on the composition of base pair fragments obtainable with specific enzyme cleavage.

An atlas of genetic maps (14) represents a compilation of linkage and restriction maps of genetically studied organisms. At the quoted price of $10 this is a most worthwhile reference item. It is available from the National Cancer Institute.

Phage Vector M13

Another coliphage which often serves as a convenient cloning vector is phage M13. It is a filamentous phage and contains single-stranded circular DNA. Its molecular weight is ~ 2 million. When it infects the *E. coli* cell it transforms into a double-stranded form which has the ability to replicate. As many as 300 copies per cell may arise. The cells which produce the phage will not undergo lysis. They continue to grow and multiply by division. The possibility of cloning single-stranded DNA from phage particles is an attractive advantage, especially for use of the single strand in nucleotide sequencing (see Chap. 8).

In contrast to phage λ, the phage M13 does not have an extended nonessential region. But through limited digestion with a restriction endonuclease a fragment of the *lac* regulatory region of *E. coli* can be cloned into M13. Also, a histidine marker may be incorporated together with a single *Eco*RI site at one end of the *his* DNA insert. As a result the modified phage M13 becomes a highly suitable vector for in vitro insertion of DNA.

Phage Injection

The process of the injection of phage DNA into a bacterial host, or, in general into a cell, is illustrated in Figure 8. The phage virion attaches itself to a specific receptor site on the bacterial cell surface. Those phages that have a tail, locate the receptor site with the tail fibers and then the tail pins penetrate into the wall. The tube has now penetrated the wall and the viral DNA is ejected through it into the cell outside of the plasma membrane. The injected DNA diffuses through the plasma membrane into the cell interior.

VIRUSES INFECTING EUKARYOTES

Any one of a wide range of viruses can infect eukaryotic entities from plant cells to higher mammals. They range from small papova and polyoma viruses (mol. wt. about 25 million) to the very large herpes and pox viruses (mol. wt. ~ 60 million). So far, only one of the small viruses, the SV40 type has become of interest in recombinant work.

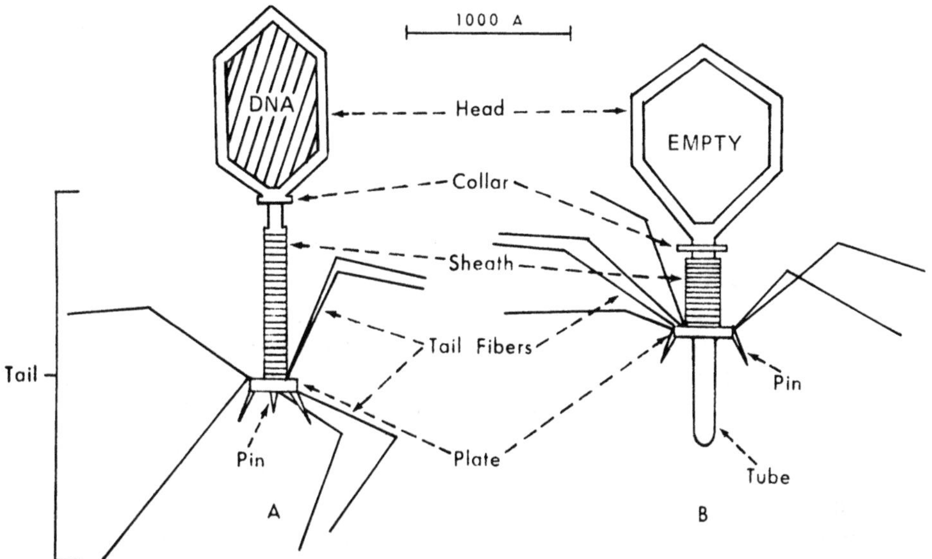

FIGURE 8 Phage injection process. *A*: Phage before injection, with full head and extended sheath. *B*: Phage after injection, with empty head and contracted sheath.

Simian Virus 40-SV40

SV40 is a small tumorigenic papovavirus which occurs in monkeys. Its genome is a covalently closed, circular DNA complex of 5226 nucleotide pairs. The virus has served as a model for the in vitro construction of mutants.

The wild type is modified, primarily by removing nucleotides at the ends, after cleavage of circular DNA to linear structure. Enzymatic trimming of the ends of the transfecting linear DNA is followed by blunt-end ligation to recyclization. Electrophoretic separation is needed to isolate the desired noninfective mutant strain.

MICROBIAL HOSTS

Remarks on Bacteria

Many years ago a conversation with a colleague brought out the versatility and tenacity of microbial life. Both of us worked in industry and we were exchanging some interesting experiences. He told about difficulties in pipeline blockages:

the lines were handling concentrated sulfuric acid. The trouble was traced to sulfur bacteria which had invaded a storage tank and had spread throughout the line system. This example is quoted to highlight the multitudinous capabilities of living organisms.

HOSTS

A microbial host is a living entity which will accept vectors, either natural or synthetic, and then proceeds to multiply and express the inserted genes. The most frequently used host is the *E. coli* bacterium. It has the ability to take up a great variety of reconstructed plasmids and phages. Another suitable prokaryote is *Bacillus subtilis*. Also, the eukaryotic yeasts have found applications. Brief descriptions of these organisms are as follows:

Escherichia coli or *E. coli* is a rod-shaped bacterium, thus a prokaryote. It is a natural component in the human colon and is normally nonpathogenic to humans. The cell structure of the bacterium is illustrated in Figure 9. *E. coli* reproduces by elongation into a longer rod while maintaining its diameter. When the length is twice that of the original dimension, the rod constricts in the middle and forms two entities, which are identical daughter cells. The so-called generation time, that is, the time interval between cell division and doubling of cell population, depends on the growth medium which is used. It can vary from about 20 minutes to several hours or days. The formation of a multitude of identical cells from a single progenitor by such growth is called *cloning.*

The *E. coli* bacterium is a Gram-negative organism. *Bacillus subtilis* or *B. subtilis*, is a Gram-positive bacterium, a prokaryote. It is a harmless species, commonly found in soil. Structurally it is a straight rod. Cell division is similar to that of *E. coli.*

Yeast is described as a unicellular, budding fungus. Fungi are eukaryotic entities. Usually, when quoted, the genus *Saccharomyces* is intended. The yeast cell is surrounded by a cell wall containing cellulose and some chitin. The cell contains a minute nucleus which is surrounded by cytoplasm. The common yeast reproduces asexually by budding, that is it forms daughter cells. The "budding" process is illustrated in Figure 10. A stimulating insight into current activities in genetic research in microbial hosts is provided in a compilation of research studies by 36 scientists (17). It is evident that genetic engineering techniques are being applied worldwide and that such activities are on the rise. For instance, many of the reported investigations make considerable use of sequencing procedures to elucidate genetic constituents. Likewise, plasmids and gene insertion therein find promising use.

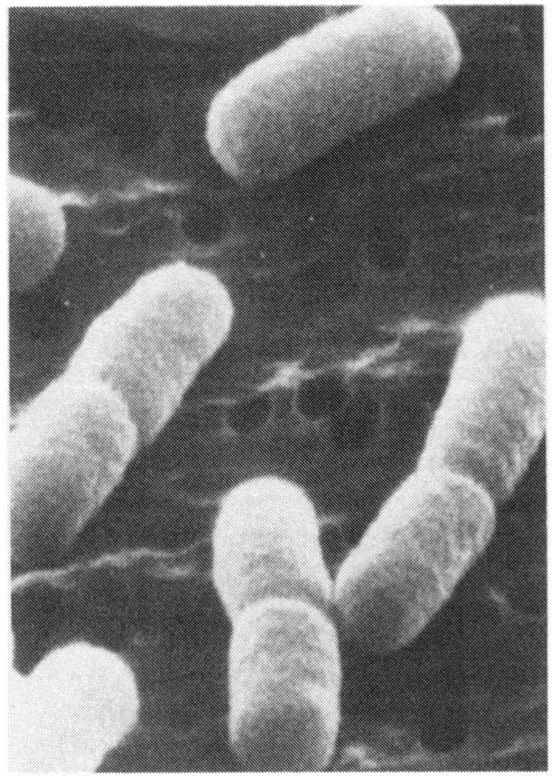

FIGURE 9 Electron Micrograph of *E. coli*. Magnification 20,000X. (Courtesy Eli Lilly & Co., Ref. 15).

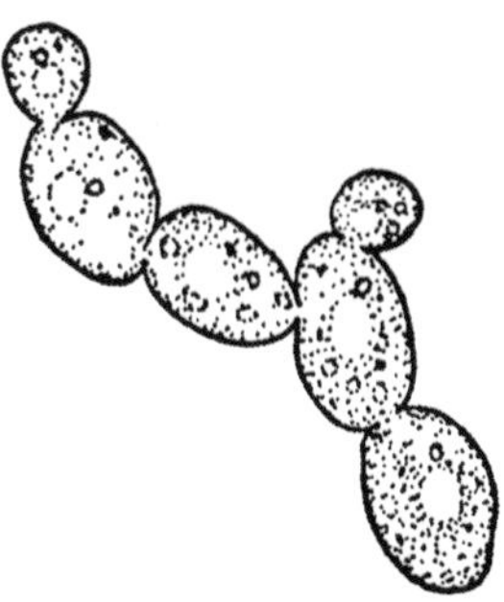

FIGURE 10 Yeast cell and budding. Chain of yeast cells from budding. (From Ref. 16.)

74-2000-00

Plasmid pBR322

30 μg	**80.00**
130 μg	**300.00**

74-2020-00

Plasmid pMB9

20 μg	**70.00**
100 μg	**250.00**

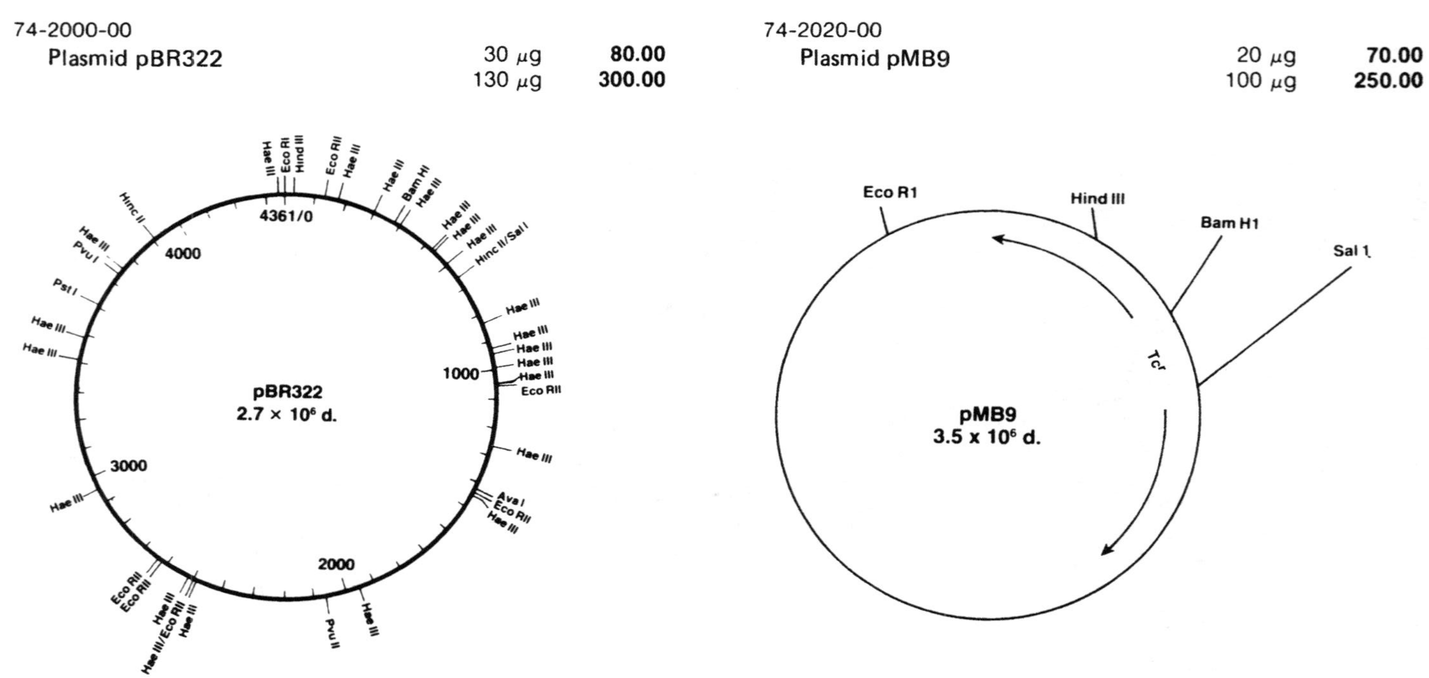

74-2010-00 Plasmid pBR325	15 μg	**80.00**
	65 μg	**300.00**
74-2030-00 Plasmid pUB110	20 μg	**80.00**
	100 μg	**300.00**

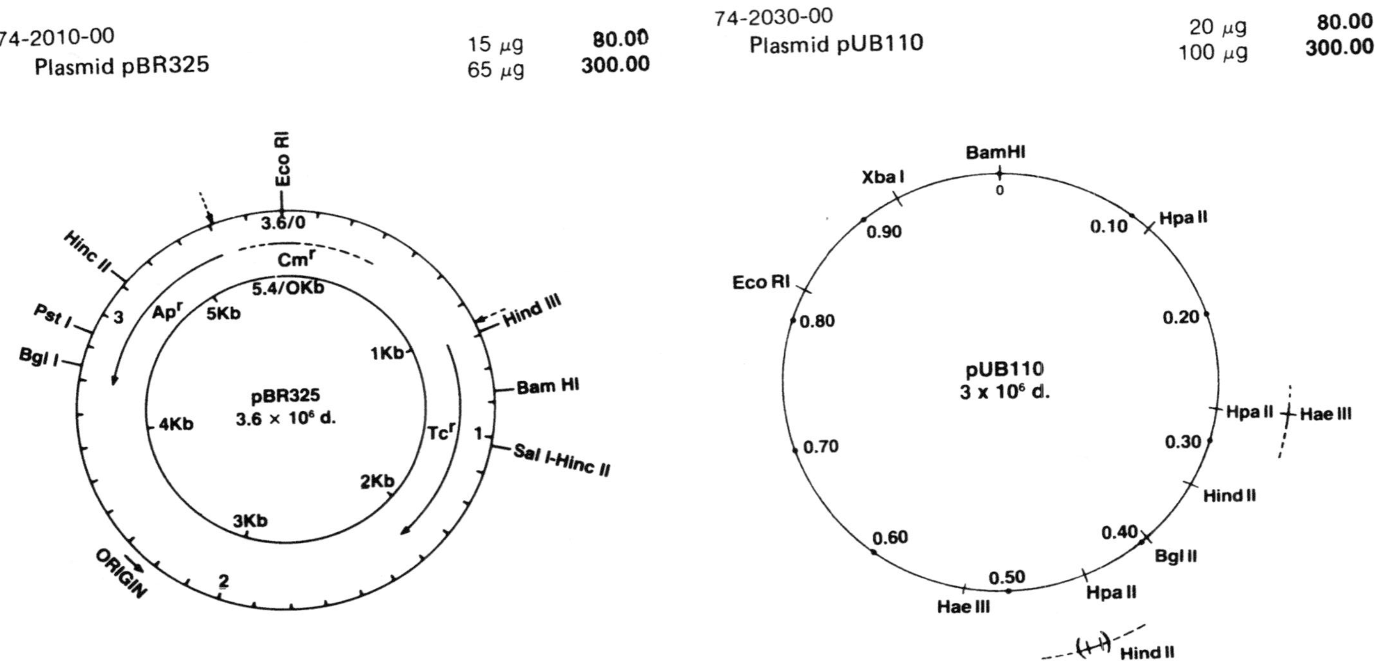

FIGURE 11 Plasmid vehicles. Four plasmid vehicles commercially available from Bethesda Research Labs are illustrated. (Reprinted by permission of Bethesda Research Laboratories, Inc., Refs. 20,21.)

Introduction into Host

The process of introducing the recombinant plasmid into a host such as *E. coli* is called "transformation." A common procedure is to use a $CaCl_2$ medium to make *E. coli* receptive to plasmid DNA. It is believed that the $CaCl_2$ modifies the cell membrane so that the vectors can penetrate into the cell. The transformation of *viral* DNA is often called "transfection," an indication that the viral DNA is infective. Cosmid DNA containing the λ cos site and recombinant DNA is introduced into *E. coli* by "in vitro packaging," which involves the in vitro assembly of the DNA into λ phage heads and creation of infectious phage particles.

AVAILABILITY OF BIOLOGICALS

The supply side of recombinant technology has reached a highly commercialized level. The selection and availability of supplies is astounding. A listing of commercial firms engaged in the supply business is beyond the scope of this chapter. A comprehensive catalog of companies through 1981 was prepared by Sittig and Noyes (18). Still, developments are accelerating at such a pace that firms new to the field are likely not to be listed. In previous chapters a number of commercial firms have been cited, but always in connection with specific developments. It is suggested that individuals who expect to become involved in the recombinant field take steps to acquire a catalog library. This can be done by perusing advertisements in trade and scientific journals. For instance a directory of biologicals can be procured from the journal *Nature* (19); it lists more than 400 companies worldwide. To illustrate the extent of availability, a page from Chemalog Hilites (20) is reproduced in Figure 11 (September 1981). The plasmids are available from Bethesda Research Laboratories (21).

REFERENCES

1. Bernard, HU, Helinski, DR. Bacterial plasmid cloning vehicles. *Genet. Eng.* 2: 133–167 (1980).
2. Hayakawa, T et al. A linear plasmid-like DNA in *Streptomyces* sp. producing lankacidin group antibiotics. *J. Gen. Appl. Microbiol.* 25 (4): 225–260 (1979).
3. Gunge, N et al. Isolation and characterization of linear deoxyribonucleic acid plasmids from *Kluyveromyces lactis* and the plasmid-associated killer character. *J. Bacteriol.* 145 (1): 382–390 (1981).
4. Hirochika, H, Sakaguchi, K. Analysis of linear plasmids isolated from *Streptomyces*: association of protein with the ends of plasmid DNA. *Plasmid* 7 (1): 59–65 (1982).
5. Colman, A et al. Rapid purification of plasmid DNAs by Hydroxyapatite chromatography. *Eur. J. Biochem.* 91(1): 303–310 (1978).

6. Vieira, J, Messing, J. The pUC plasmids, an M13mp7-derived system for insertion mutagenesis and sequencing with synthetic universal primers. *Gene* 19 (3): 259–268 (1982).
7. Sanger, F, Nicklen, S, Coulson, AR. DNA sequencing with chain-terminating inhibitors. *Proc. Natl. Acad. Sci.* (*USA*) *74* (12): 5463–5467 (1977).
8. Messing, J, Crea, R, Seeburg, PH. A system for shotgun sequencing. *Nucl. Acids Res.* 9 (2): 309–321 (1981).
9. Clewell, DB, Helinski, D. Supercoiled circular DNA-protein complex in *E. coli*: Purification and induced conversion to an open circular DNA form. *Proc. Natl. Acad. Sci.* (*USA*) 62 (4): 1159–1166 (1969).
10. *Sephacryl.* Bulletin, Pharmacia Fine Chemicals AB, Uppsala, Sweden (Piscataway, NJ); also *Gel Filtration* bulletin.
11. Davis, BD et al. (1973). *Microbiology Including Immunology and Molecular Genetics*, 2nd Ed., Harper and Row, Hagerstown, MD.
12. Simons, K et al. How an animal virus gets into and out of its host cell. *Sci. Am.* 246 (2): 58–66 (1982).
13. Williams, BG, Blattner, FR. (1989). Bacteriophage lambda vectors for DNA cloning. In *Genetic Engineering, Principles and Methods*, Vol. 2. 201–281, J. K. Setlow, JK, and Hollaender, A (Eds), Plenum Press, New York.
14. O'Brien, SJ. (Ed). (1980). *Genetic Map.* Building 560, Room 11-85, National Cancer Institute, Frederick, Md 21701.
15. *Recombinant DNA and Biosynthetic Human Insulin.* Eli Lilly and Co., March 1981. Prepared by John Dolan-Heitlinger.
16. Alexopoulus, CJ, Mims, ChW. (1979). *Introductory Mycology*, 3rd Ed., John Wiley & Sons, Inc, New York.
17. "*Molecular Genetics in Yeast*," Eds.D. Von Wettstein, *et al, Proc. of Alfred Benson Symp.*, Royal Dutch Acad. Sci. and Letters, Publ., Munksgard, Copenhagen, Distributor: Nankodu, Tokyo (1981).
18. Sittig, M, Noyes, R. *Genetic Engineering and Biotechnology Firms, U.S.A.-1981.* Sittig and Noyes, 84 Main St., P.O. Box 75, Kingston, NJ 08528.
19. *Nature, Directory of Biologicals 1982. Nature* 15 East 26th Str., New York, NY 10010 (Quoted at $45).
20. *Chemalog Hilites.* (1981). Chemical Dynamics Corp., South Plainfield, NJ.
21. Bethesda Research Laboratories, Gaithersburg, MD.

7
Recombinant Techniques

Genetic engineering proper comes into play when structural changes in DNA, and, therefore, gene manipulations are carried out by human intervention. Changes in cells, that is *mutations*, occur naturally at all times. Such events are haphazard and can result in either beneficial or damaging effects to the organism. Frequently, they are of a minor nature and essentially benign. The prevailing opinion is that every mammalian body, including the human body, carries a number of gene defects which, however, are entirely inconsequential and do not affect the normal process of life.

The interventional procedures to create changes in DNA, that is, create *recombinant* DNA, are conducted to prepare organisms to carry out a useful function. Although the present state of the development is already remarkably advanced, the work is essentially limited to the manipulation of lower organisms. Nevertheless, recombinant DNA technology has been called the "ultimate technology" (1).

BASIC CONCEPTS

Real genetic engineering became possible in 1970. Present day DNA technology is completely dependent upon the ability to cleave DNA at specific locations with so called *restriction endonucleases*, which are enzymes (see Chap. 5).

Conceptually, the alteration of DNA, which in essence is a chemical compound, a very large polymer molecule, is quite simple. The molecule is cleaved at specific locations and then reassembled (ligated) with another DNA segment to form a new entity (hybrid formation). The recombined DNA structure then imparts *phenotypical* properties to the cell acquired from the new gene expressions. The actual physical and biological procedures needed to accomplish such gene transfers are not exceedingly complex, but require a considerable degree of knowledge and manipulative skill. There is already an impressive amount of information, which permits the formulation of a specific procedure to accomplish a desired result.

GENERALIZED SCHEME

DNA splitting and gene insertion are presently conducted by the use of *prokaryotic* entities. Usually, a plasmid or a phage is used as the *vector* to receive gene insertion. The relatively simple circular prokaryotic DNA is cleaved at a specific location by means of suitable restriction enzyme. Some other DNA, having the desired gene character, also cut-to-size, is added to the vector in an aqueous medium containing appropriate chemicals and a ligase, an enzyme which promotes recombination, is added to join the DNA strands, thus forming a modified circularized DNA. This new *chimera,* plasmid or phage, is introduced into a bacterium where it is able to replicate and produce a desired product through its gene character. The procedures are illustrated in a rather simplified format in Figures 1 and 2.

If a biochemist knowledgeable in recombinant techniques were to sketch a diagram to illustrate the basic recombinant concepts without consulting any reference sources, the drawing would look like Figure 3. Very simply one would select a plasmid, perhaps a pBR322 which has ampicillin and tetracycline resistance genes, cleave it with *Bam*HI in order to inactivate the Tet^R gene and generate an open structure (one could envision it as a linear fragment) with sticky ends. Similarly, one would cleave a suitable donor DNA, perhaps also with *Bam*HI to create a number of fragments which also would have sticky ends. From the collection of fragments one isolates a desired sequence by gel separation, excision of the sequence, and elution of the specific DNA fragment.

It is a simple matter to ligate the opened plasmid and the DNA sequence. The reaction can obviously result in three types of recombinant structures:

- A recircularized plasmid having Amp^R and Tet^R
- A recombinant plasmid containing the desired combination of plasmid fragment with Amp^R only and the donor DNA
- A circularized donor DNA fragment resulting from the linking up at the sticky ends

Introduction of the mixed recombinants into a cloning vehicle such as *Escherichia coli* will give a mixed population of clones. These then can be screened with successive platings on Amp- and Tet-containing culture media.

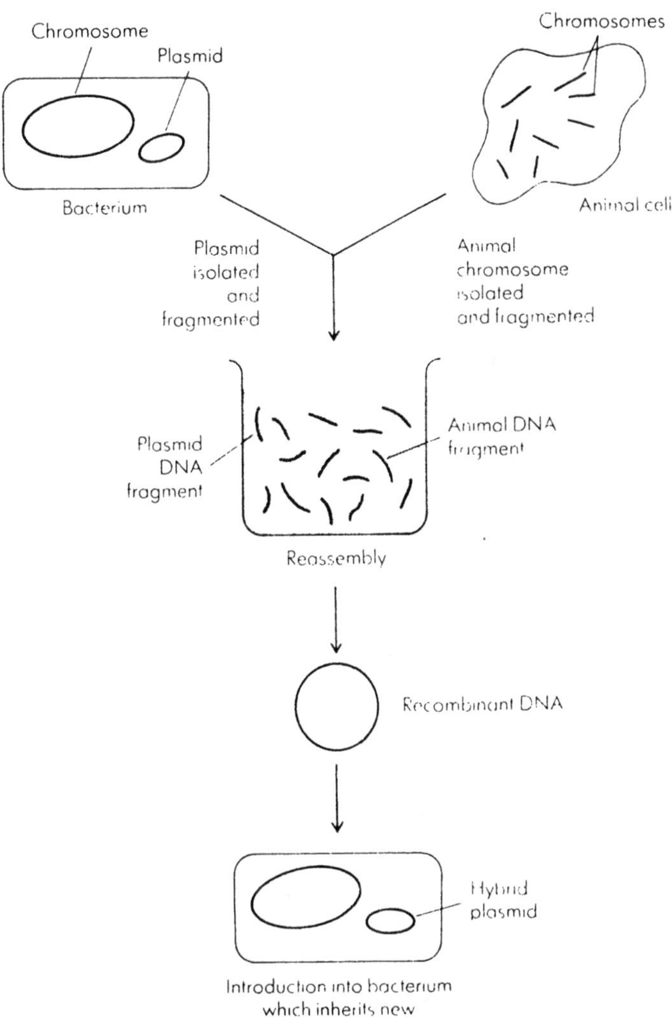

FIGURE 1 Schematic of recombinant procedure. Plasmid in bacterium is isolated and linearized, combined in solution with foreign DNA, recircularized and introduced into a bacterial host. (Courtesy McGraw-Hill, Ref. 2).

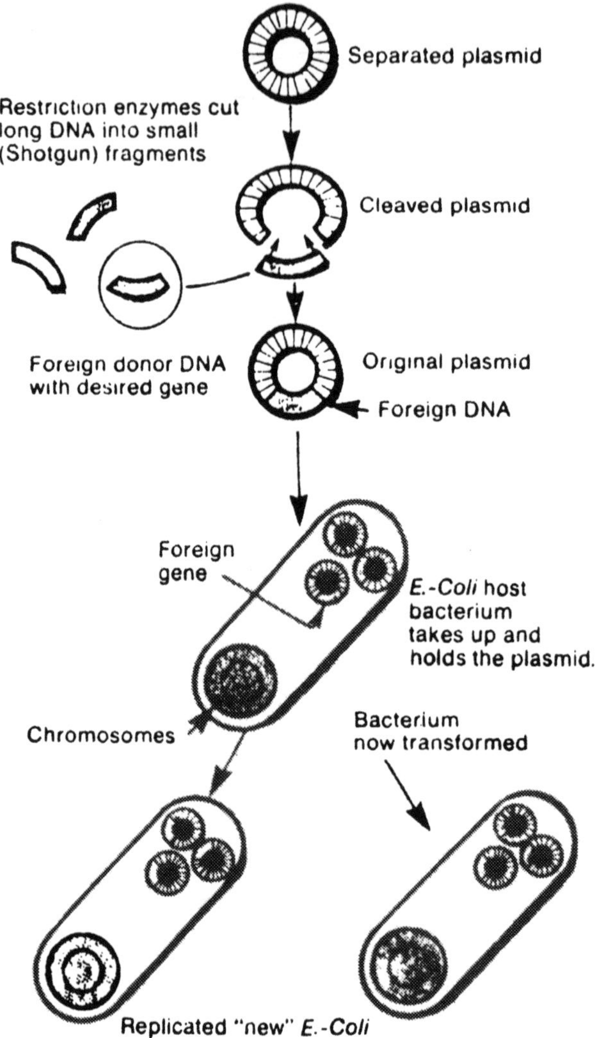

FIGURE 2 Schematic DNA insertion into plasmid. Plasmid is cleaved and foreign DNA is inserted, recircularized and cloned in host. (Courtesy Food Engineering, Ref. 3).

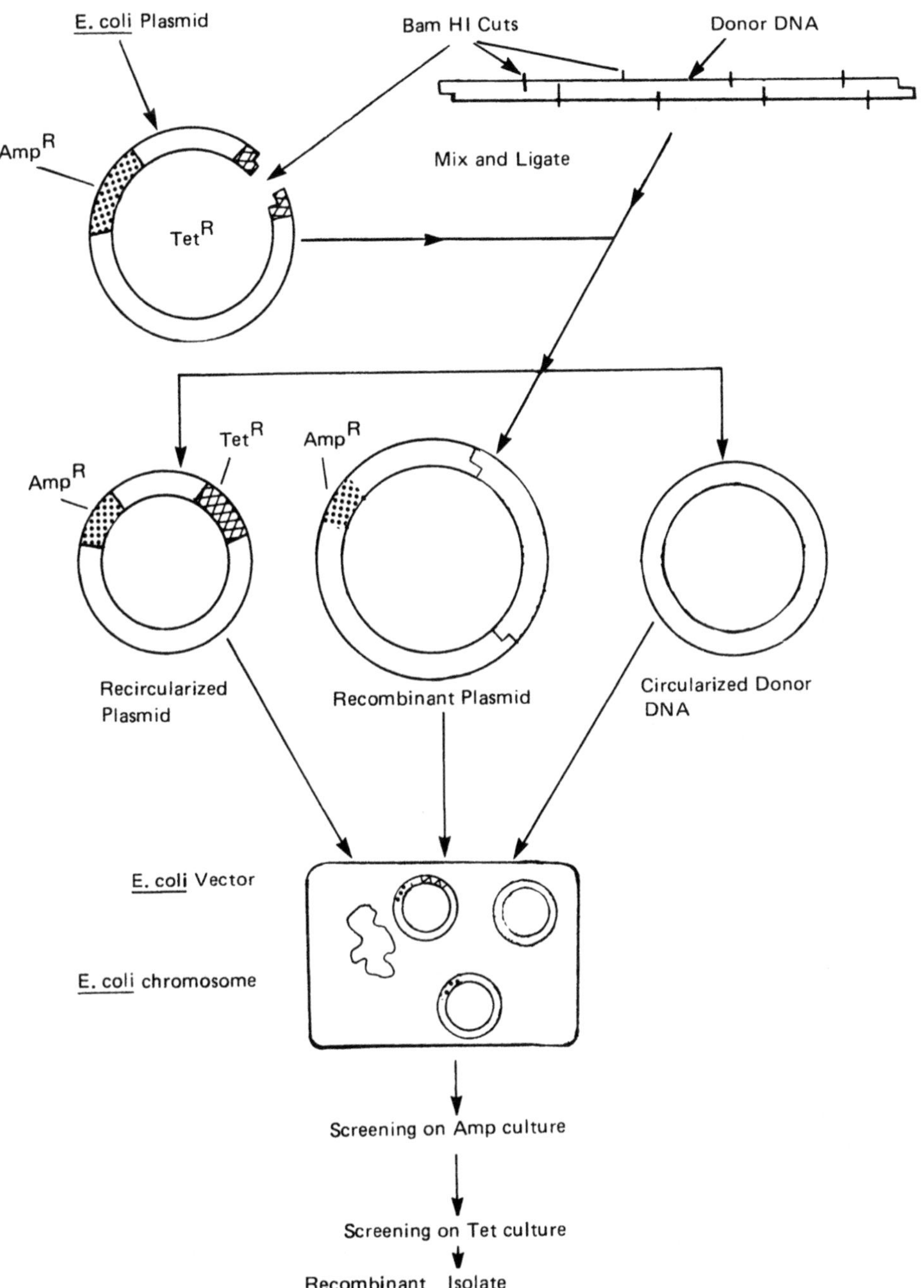

FIGURE 3 Generalized foreign gene insertion. Individual steps to make plasmid receptive to gene insertion, and antibiotic screening for selecting properly transformed cells.

The final selection will be the new recombinant which will grow on the Amp culture but not on a Tet medium.

CUTTING SITES OF VECTORS

An overview of available vectors, such as plasmids and phages, was presented in Chapter 6. To appreciate the potential intricacies in recombinant procedures, a somewhat more detailed treatise of cleaving characteristics is indicated. A generalized summary of commonly used plasmid vectors (Table 1) shows the sources (original host) and the hosts that will accept the respective plasmids. As previously indicated, the most widely used prokaryote is *E. coli*, both as a supplier of plasmids and as a host recipient.

TABLE 1 Plasmid Vector Systems

Vector	Original host	Host range
Gram-negative bacteria		
Col E1	*Escherichia coli*	*E. coli*
pBR322	*E. coli*	*E. coli*
RSF1010	*E. coli*	*E. coli*, some *Pseudomonas* spp.
pRK290	*Klebsiella aerogenes*	Most gram-negative bacteria
RP1	*Pseudomonas aeruginosa*	Most gram-negative bacteria
Gram-positive bacteria		
pUB110	*Bacillus*	*Bacillus* spp.
SCP-2	*Streptomyces*	Some *Streptomyces* spp.
Fungi		
YEp; YRp	*Saccharomyces*	Yeast
Plant Cells		
Ti	*Agrobacterium tumefaciens*	Dicotyledenous plant species
Animal cells		
Defective SV40[a]	Mammalian cells	Mammalian cell lines

[a]Control of transmittance is obtained in part by use of defective viral forms

Source: Reprinted from Food Technology, *35* (1), 29 (1981). Copyright © by Institute of Food Technologists.

Bacillus subtilis is the only other bacterium that has been studied intensively. Much success has been achieved with plasmids from *Staphylococcus aureus* when introduced into *B. subtilis*. Thus, plasmid DNA from *S. aureus* is readily taken up by *B. subtilis*, and is capable of autonomous replication.

Some of these plasmids are temperature sensitive and segregation takes place above ~ 35°C. The antibiotic resistances are: Cm-chloramphenicol, Em-erythromycin, Sm-streptomycin, Km-kanamycin, Tc-tetracycline. An advantage of using *B. subtilis* is that it has been processed in large-scale fermentation procedures and thus adequate methods of handling are available. Also it is very safe and presents no biohazard.

Specific Procedures

Manipulations involved in recombinant work concern the behavior of DNA under a variety of experimental conditions. A double-stranded DNA (ds-DNA) can be converted to single strands, the ds-DNA can be cut at desired positions, foreign DNA can be inserted in the linear configuration, the ends of linear DNA can be tailored, and recircularization can be conducted. These procedures will be treated in the subsequent discussion.

DNA DENATURATION AND RENATURATION

Denaturation

The chains of the DNA duplex come apart when the hydrogen bonds between the bases break. This can be accomplished as follows:

1. In solution: increasing the temperature.
2. In solution: titrating with acid or alkali; use of acid protonates, the ring N atoms of G and C; use of alkali deprotonates the ring N atoms of G and T. (Use of acid is undesirable as it attacks the purine glycosidic bonds.)

Stability of the DNA duplex is a function of the number of triple-hydrogen-bonded GC pairs; a higher temperature is needed or higher pH when mole fraction of GC is high. Separation, or melting (thermal) is monitored by an increase in absorbance (hyperchromic effect).

Renaturation or Annealing

Even after complete separation of the two chains, a recombination or a matching up of complementary single strands can be accomplished with ease. All that is necessary is to incubate the mixture at a temperature about 25°C. below the T_m (= midpoint melting temperature) whcih depdnds on the GC content. The strands of complementary composition then approach and blend together side by side. (*Note*. combining end-to-end is ligation.)

CLEAVING OF DNA

Synthetic intervention in the creation of recombinant DNA first requires cutting a nucleotide sequence at a predetermined site. This can only be accomplished by using a specific restriction endonuclease–RT a restriction enzyme. Obviously, this calls for a knowledge of specificity of RT action based on a fund of information acquired over many years of investigation.

As a specific example, the effect of the endonuclease which cleave the *E. coli* plasmid pBR322 is illustrated by Figure 4 and Table 2. The total base pair (bp) composition of the plasmid is 4362. Counting starts at the *Eco*RI site. The indicated enzymes cut at the recorded locations in the figure. From the table, one can then ascertain the specific cutting sites in the polymer chain. For instance, *Eco*RI will cut the sequence G↓AATTC, that is after the guanine G base, or *Pst*I will cut after A in the sequence CTGA↓G. So, the *Pst*I cleaving procedure will look as follows:

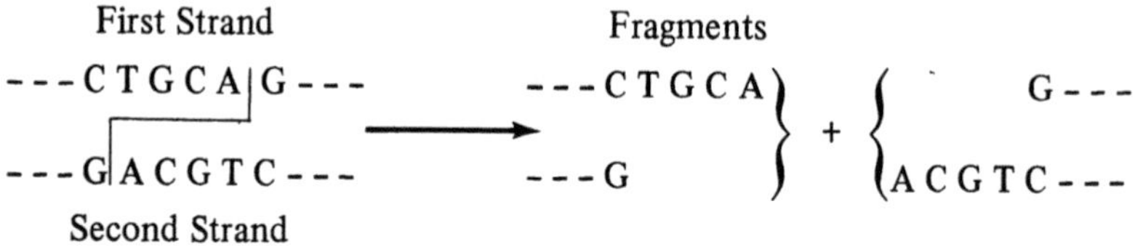

```
   First Strand                          Fragments
---C T G C A|G---                  ---C T G C A)       (         G---
            |                                   } + {
---G|A C G T C---      ------->    ---G         )       (A C G T C---
   Second Strand
```

The result is the creation of fragments with overlapping ends. The nucleotide base pairs of pBR322, at which the following enyzmes cleave the plasmid are at:

*Eco*RI	1
*Hind*III	29
*Bam*HI	375
*Sal*I	650
*Pst*I	3641
Total	4362

The curved arrows, inside of the second strand, indicate the extent of the mRNA coding region for tetracycline (clockwise) and for ampicillin mRNA

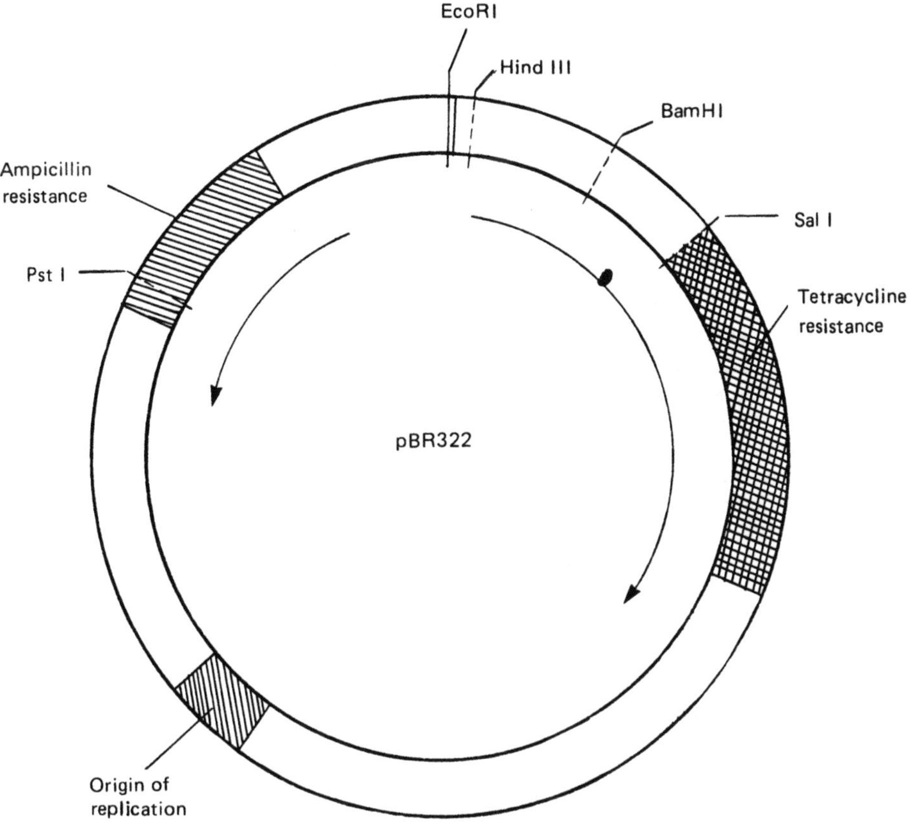

FIGURE 4 Simplified structural map of plasmid pBR 322. The location of the five specific cleaving sites are shown. Enzymes listed make single cuts.

(counterclockwise). When an enzyme makes a cut in the nucleotide sequence within the antibiotic resistance regions, as for instance *Sal*I in Tet and *Pst*I in Amp, the resistance property becomes inactivated.

The reactivation conditions to accomplish cutting call for optimal incubation of plasmid and enzyme. Different enzymes require variations in the aqueous medium in factors such as ionic strength (salt content), pH value, and temperature to achieve optimal activity. The use of the *Eco*RI enzyme requires:

10 m*M* Tris HCl
7.4 pH
50 m*M* NaCl

TABLE 2 Target Sites of Some Restriction Endonucleases

Bacillus amyloliquefaciens H	*Bam*HI	G↓GATCC
Escherichia coli RY13	*Eco*RI	G↓AA*TTC
Haemophilus aegyptius	*Hae*III	GG↓C*C
Haemophilus influenzae Rd	*Hind*II	GT(T over C)↓(G over A)A*C
Haemophilus influenzae	*Hind*III	A*↓AGCTT
Providencia stuartii	*Pst*I	CTGCA↓G
Streptomyces albus G	*Sal*I	G↓TCGAC
Streptomyces stanford	*Sst*I	GAGCT↓C

Notation: Recognition sequences from 5′ to 3′ single strand only. C over (T) means C could occupy T location. A* is N^6-methyladenine, C* is 5-methylcytosine. *Explanation*: Example *Eco*RI cuts at indicated position: that is it recognizes a specific base sequence:
(5′ phosphate end) 5′- GAA TTC - Palindromic
3′- CTT AAG - Sequence
Axis of symmetry

6 m*M* β-mercaptoethanol (stabilizes sulfhydryl groups and reduces disulfide linkages in the enzymes, for instance in cystine)
100 μ g/ml gelatin or bovine serum albumin

CREATION OF COHESIVE ENDS

*Eco*RI digestion makes single-strand cuts at four nucleotide spacing in the complementary strand of its target sequence. This generates fragments with protruding 5′ ends, that is, cohesive ends (Fig. 5).

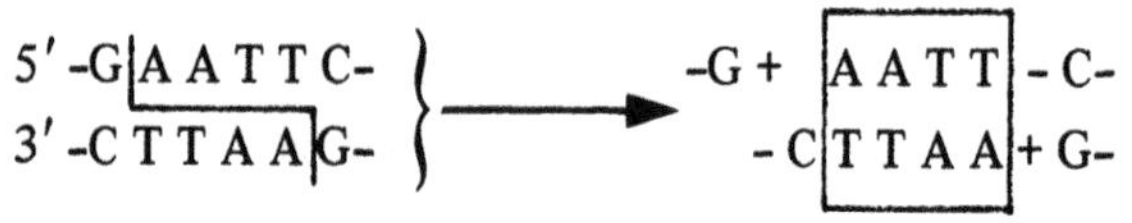

```
5′-G|A A T T C-  }          -G +  |A A T T| - C-
3′-C T T A A|G-  }  ----->   - C|T T A A| + G-
```

Upon hydrogen bonding the overlapping 5′ ends can either form a linearized aggregate or circularize (Fig. 5). This *Eco*RI specificity is found with many Type II restriction endonucleases (different specificities recognized) and is very useful in a multitude of recombinant procedures.

In most cases it is not known which restriction enzyme can be used to isolate the desired gene as a complete entity. If the wrong enzyme is used it may cleave in the middle of the coding sequence. For this reason, when investigators clone unknown DNA sequences they only partially digest the foreign DNA with the restriction enzyme. Partial digestion can be achieved most easily by limiting the amount of enzyme used and by allowing the reaction to proceed for only short periods of time. Partial digestion ensures that there will be a fragment within the population being cloned that will contain the complete DNA sequence of the gene, even if that gene sequence contains a site for the restriction enzyme used for cloning within it. A second method to ensure the presence of the complete gene sequence is to shear the DNA, generating random fragments with blunt ends. After isolating the fragments of the desired size by electrophoretic or gradient centrifugation separation, the fragments can be joined by blunt end ligation.

Gene Insertion

The incorporation of foreign DNA into prokaryotic and eukaryotic cells is the basic problem in gene manipulation. A molecule which can replicate DNA

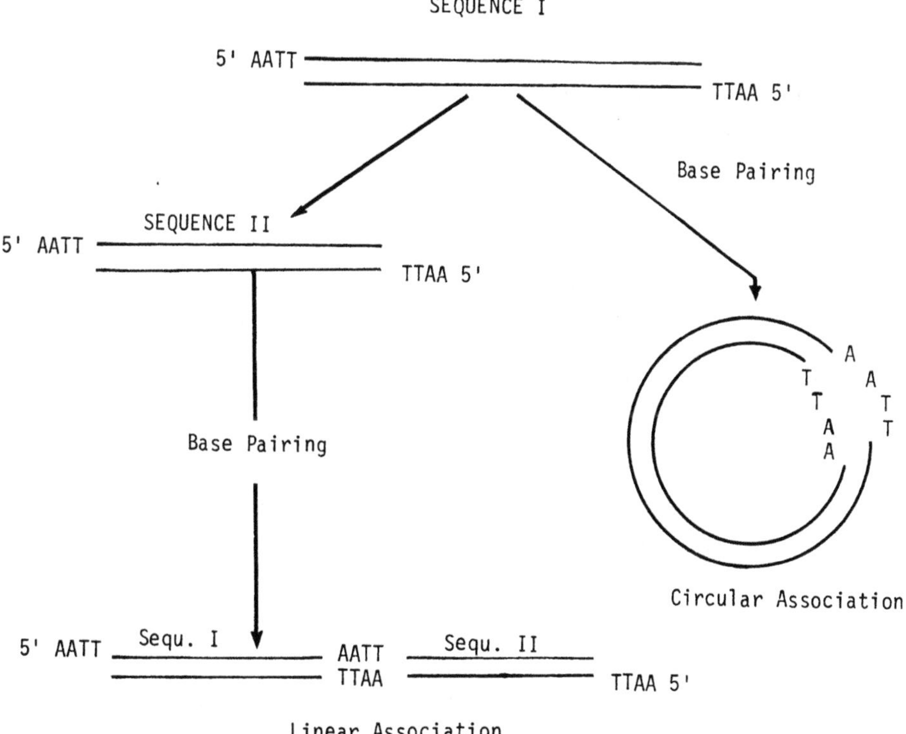

FIGURE 5 Creation and joining of cohesive ends. Cleaving DNA with *Eco*RI produces overlapping complementary sequences. Linear association also called intermolecular linkage. Circular association is considered intramolecular linkage.

because there is contained in it an "origin of replication" is called a replicon. In bacteria and viruses there is usually only one such structure per genome. Ordinary fragments of DNA are normally not replicons and the DNA characteristic may be lost from the host cell in the process of cell multiplication (gradual dilution).

In the introduction of DNA fragments into a suitable structure, for instance, a bacterium, one does not want to be restricted to the use of replicons only (they may not contain the desired characteristic) and then a means has to be used to utilize whatever DNA fragment is needed. This is accomplished by attaching the desired fragments to suitable replicons (also called *vectors* or *cloning* agents). After the vector or the DNA to be inserted is tailored by cleaving to contain the desired and needed sequences, it must be assembled by recombining.

RECOMBINING DNA FRAGMENTS

The available methods for joining DNA segments are as follows:

1. *DNA ligase* catalyses the covalent linkage of annealed *cohesive ends* (CCC) resulting from the cleaving action of specific restriction enzymes. DNA ligase is produced by *E. coli* and *phage T4*. This enzyme seals nicks on single strands between adjacent nucleotides in a double-stranded DNA configuration. To be active, the *E. coli* enzyme needs nicotinamide adenine dinucleotide, *NAD* (a cofactor).

The conventional presentation of DNA is schematic diagrams in that of two parallel lines or two concentric circles. Each line indicates one of the spirals of the DNA configuration (Fig. 5). The action of the *DNA ligase* is illustrated in Figure 6. The restriction enzymes used for cleaving, such as *Eco*RI, create single strand cleavages in the foreign DNA, so that at each end there is a gap of four nucleotides, namely the AATT and TTAA at each 5′ end of the DNA. These are the protruding 5′-*termini*. They are also called sticky or *cohesive* ends. The plasmid vector is opened to contain complementary nucleotides (AATT) on the exposed ends. Then, the fragments can undergo hydrogen bonding as the ends are complementary, that is: A- -T, T- -A, etc. Through the annealing process (a side-by-side orientation of the duplex strands), the foreign DNA is incorporated into the opened plasmid. When this occurs, unsealed ends remain at the insertion loci. These are then "sealed" by means of DNA ligase. Subsequently, the newly created plasmids are screened to eliminate plasmids which have not been altered. The desired viable plasmids, the recombinant forms, are introduced into a suitable bacterial host for cloning.

2. DNA ligase from phage T4-infected *E. coli* will catalyze the joining of *blunt ends* (also called flush ends) by forming phosphodiester bonds. The T4 enzyme requires adenine triphosphate [(*ATP*), a cofactor)]. The terms "nick"

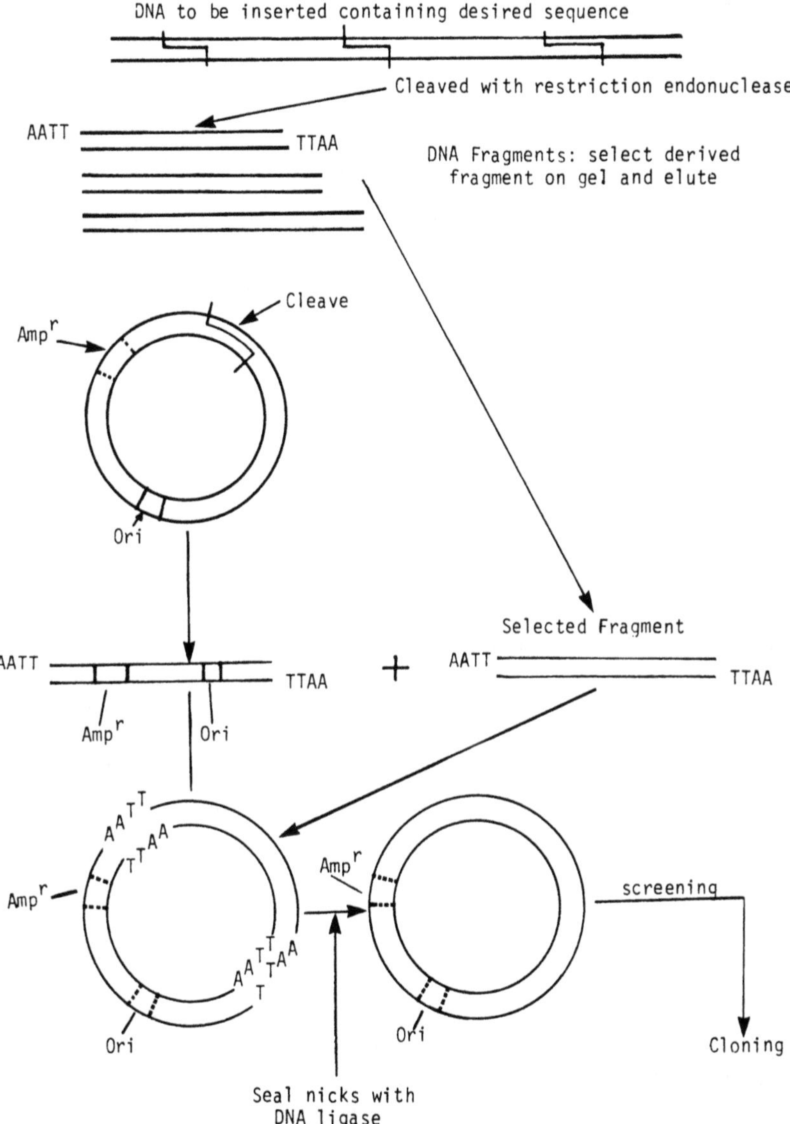

FIGURE 6 Use of DNA ligase in cleaving and recombining of DNA. Demonstration of covalent DNA recombination through overlapping termini created by *Eco*RI.

and "cut" are used in describing cleaving procedures. *Nick* means cleavage of one strand. *Cut* means cleavage of *both* strands. Joining with DNA ligase (T4 phage): Some enzymes, as for instance *Hae*III (Table 2) cleave both DNA strands all the way through and so generate blunt ends at a

```
GG | CC
CC ▼GG
```

sequence. To ease recombination in later steps a so-called *linker molecule* is employed. A linker is a palindromic oligonucleotide fragment which will form the so-called sticky ends when it is ligated to a blunt ended DNA. The linked DNA then is ready for insertion into a vector.

Frequently, a decameric linker molecule is cleaved with *Eco*RI, so that:

```
G C G▼A A T T C G C
C G C T T A A▲G C G
```

and the palindromic AATTC sequences ligate to the blunt end cuts. A rather elegant visualization of the use of linkers and also adapters is presented in Figure 7 (5) at the top. As the linker has an *Eco*RI site, the ends of the linear combination can be cleaved to give TTAA ends. Then, insertion of the tailed foreign DNA into an appropriate plasmid vector will proceed with ease.

An *adapter* is an oligonucleotide fragment containing recognition sites for more than one restriction endonuclease. By ligating an adapter to a plasmid or other vector, this vector may be "adapted" to recognize more than one restriction enzyme at the site in which the adapter is inserted (see Fig. 7, bottom).

3. Homopolymeric 3′ single-stranded end tails can be synthesized by use of the enzyme "terminal deoxynucleotidyl transferase," obtained from calf thymus. The aim is to create a circular plasmid type, configuration from linear DNAs only.

Homopolymer Tailing

The attachment of multiple nucleotides, that is oligo (dA) and oligo (dT) structures to blund-ended DNA is called homopolymer tailing. Figure 8 shows the addition of -AAA- - and -TTT- - ends to the respective 3′ ends of the DNA. The exposed 3′-OH groups have been prepared by pretreatment with phage λ exonuclease or RT *Pst*I (table 2). Once the homopolymer tails are in place circular closure will readily take place.

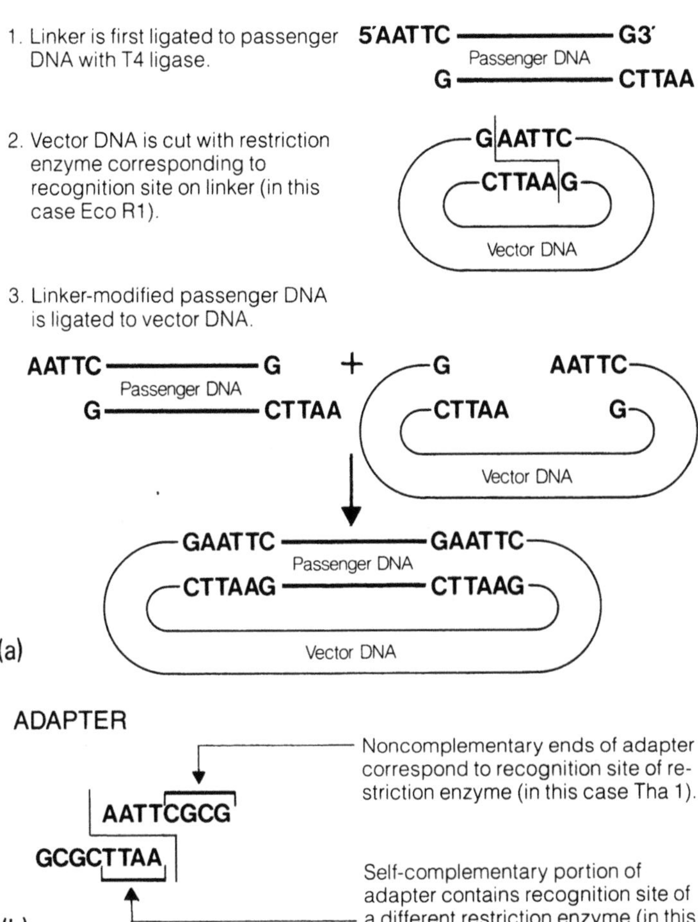

FIGURE 7 Use of linkers and adapters. (a) Use of linkers to convert a blunt end sequence to a sticky end configuration. (b) Use of adapters. (Courtesy Biosearch, San Rafael, CA 94901, Ref. 5).

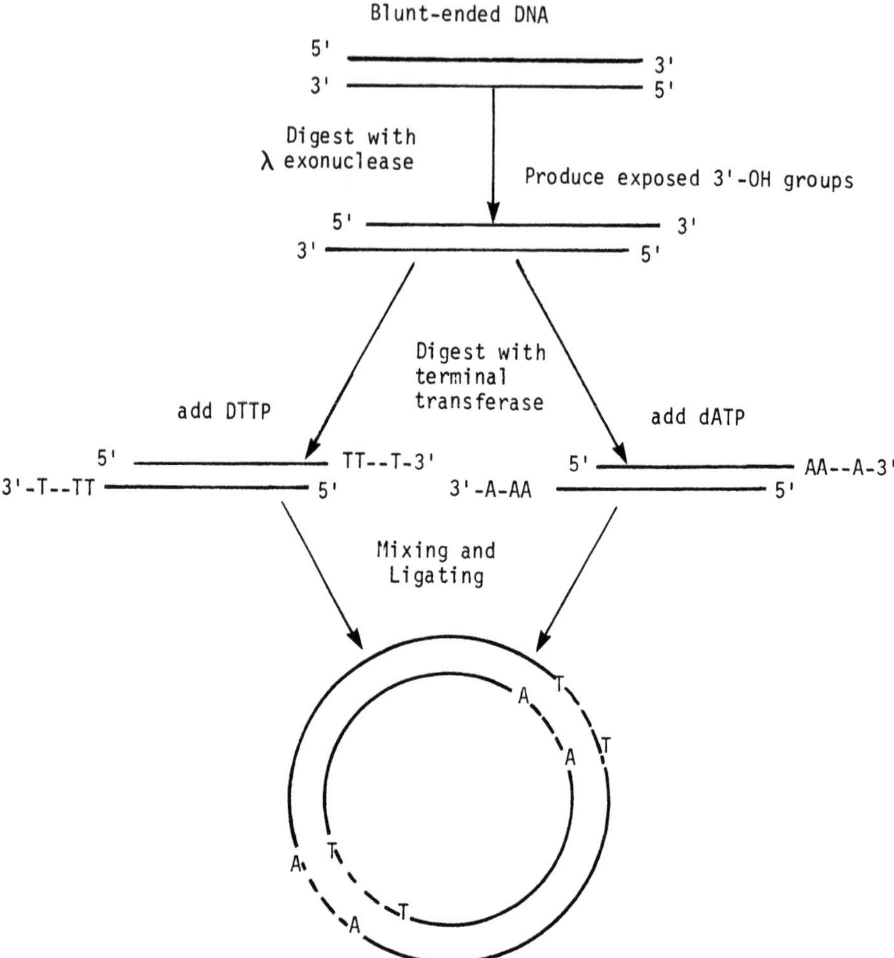

FIGURE 8 Homopolymer tailing. Addition of oligo (dA) and (dT) tails to linear DNAs permits circularization of cohesive end fragments.

BACTERIAL HOSTS CLONING VEHICLES

The introduction of the tailored vector plasmid (or phage) DNA into a "host" vehicle can be accomplished by "*transformation*," which is defined as "the introduction of an exogenous DNA preparation (also called transforming DNA) into a cell." The cells, usually bacteria, which have taken up plasmids are called "transformants."

A considerable number of bacterial and fungal (yeast) hosts as well as some plant cells and animal viruses have been used for insertion of DNA structures.

Plasmid Cloning Vehicles

A myriad of plasmids have proved to be suitable vectors for gene insertions. Because of easy availability and safety in handling, plasmids contained in *E. coli* bacteria have received the most attention. Bernard and Helinski published a comprehensive treatise on plasmids (6). Information on cloning vehicles from *E. coli, S. aureus,* and some in vitro construction is given in Appendix IV (7). The data cover replicon type, size in kilobase pairs, selective markers (antibiotic resistance), and enzymes which cut at single restriction sites. The original article also contains a discussion of the significance of the *lac* promoter in the plasmid gene sequence where it exercises transcriptional control. Also, data on plasmids with *lac* promoters and λ phage promoters are given. All in all, the voluminous information makes this publication a particularly valuable reference source.

CLONING

A basic requirement for cloning is the presence of an "origin of replication" in the DNA molecule which is to be replicated. In bacteria and viruses there is normally only one such sequence in each genome. That structural combination is called a *replicon*. Usually, DNA fragments, which are to be cloned for the specific gene that they carry, do not contain a replicon. At this point the recombinant procedure comes into play. The DNA fragment is inserted into a plasmid or phage, as these entities contain their own replicons and so become "DNA carriers" or vectors. The engineered vectors are then inserted into a microbial host which recognizes the vector's replicon. From here on the cloning procedure begins, that is the formation of new microbial entities which are identical to the original structure. Some foreign DNA fragments of course, can contain replicons which may be recognized by the host.

THE CLONING PROCESS

Gene Introduction. The insertion of a specific DNA sequence into a vector, that is, the carrier of the new gene, requires the following conditions (6):

1. A DNA vehicle (vector or replicon) which can replicate after foreign DNA is inserted in it.
2. A DNA molecule to be replicated: the foreign DNA insert, the passenger.
3. A method for introducing a passenger into the vehicle.
4. A means for introducing the vehicle carrying the passenger into a host organism in which it can replicate; DNA transformation or transfection (the term transfection is used frequently with viral DNA).
5. A means for screening or genetic selection for those cells that have replicated the desired recombinant molecules. This screening or selection process provides a way to recover the specific recombinant DNA in pure form. (The screening process by means of antibiotic resistances was illustrated in Figure 3.)

Elaboration:

1. Cloning vehicles. Frequently used vehicles are bacterial plasmids from *E. coli* and *B. subtilis*; also used are yeast, streptomyces, bacteriophage λ, phage M13, and SV40 virus.
2. DNA to be replicated. DNA fragments having the desired genome can be prepared by mechanical chopping (nonspecific fragments) or by digestion with restriction endonuclease (specific site cleavage). For a particular gene selection, the fragmented DNA is separated by means of gel electrophoresis (e.g., agarose gel). The fractions of the desired size are excised from the gel slab and eluted.

   Chemical synthesis of short segments (~ 40 bp) of oligodeoxynucleotides has been developed. Enzymatic joining of such segments to form double-stranded DNA containing synthetic genes is possible (see section on Synthesis). Thus, a number of desirable genes can be synthesized.
3. Joining Methods. Restriction enzymes which produce staggered cuts (sticky ends) can make single-strand termini of 2 to 4 nucleotides. These can be annealed to DNA from other sources and ligated to form hybrid DNA molecules. If recircularizing of the vector must be avoided, treatment with alkaline phosphatase and ligating to the target DNA can be used. This is especially useful in the cloning of some cDNA molecules.

   If blunt-end cleavage occurs, cohesive ends can be added by ligating synthetic linkers or adaptors. The linkers are duplex DNA molecules which contain the recognition sequence for a restriction endonuclease that then produces cohesive termini. The linkers themselves are blunt-end to match up with the blunt-end cleavage. Linkers and adaptors are available commercially.

   Another method is the use of complementary homopolymeric tails at the termini of the vehicle and the passenger DNA with the help of deoxynucleotidal terminal transferase and deoxynucleoside triphosphates.

4. Introducing and Screening for Recombinant DNA in a Host. The preferred host at present, still is *E. coli* K12. The DNA is introduced into the host by transformation or transfection. Introducing recombinant plasmids into *E. coli* generally uses the calcium ($CaCl_2$) treatment procedure of Cohen (8). The bacteria containing the plasmids can be selected by using drug resistance marker genes on the vector DNA. Thus, the bacteria that have taken up the plasmid, which carries a drug resistance, (e.g., Amp), are plated on a nutrient containing Amp. Then, the only bacteria which will grow are those which have taken in the plasmids.

   Another procedure uses readioactively labeled nucleic acid probes which are hybridized with the DNA carrying the foreign gene. This requires electrophoretic separation and excising the radioactive sites from the gel.

SPECIFIC CLONES

Eukaryotic DNA for Cloning

To prepare a eukaryotic DNA for gene cloning, the preferred method is to isolate the eukaryotic mRNA and obtain from it a cDNA by reverse transcription. By using the mRNA, a structure is available in which the noncoding DNA sequences have already been excised (see Chap. 4).

Fortunately, most of the eukaryotic mRNA sequences will be in a polyadenylated form, that is, they will contain a large sequence of adenylates at the 3′ terminus. This permits the creation of a primer by adding short oligo (dT) molecules. These readily hybridize with the poly (dA) sequence. The presence of a primer is essential for the reverse transcription process.

The creation of double-stranded cDNA from a suitable mRNA template is pictured conventionally (9-11) as shown in Figure 9. mRNA, having only adenines at the 3′ (free 3′-OH group), is used to hybridize short oligo (dT) with the poly (A) sequence. This provides a *primer*. Reverse transcription is conducted with reverse transcriptase giving an RNA-DNA hybrid. Alkaline hydrolysis leaves the alkali-resistant cDNA. Conversion to ds-DNA with polymerase gives a hairpin structure, which can be opened with S1 nuclease. Tailing with oligo (dT) gives the overlapping ends which then match up with a poly A-tailed linearized plasmid.

Once the cDNA has been prepared and isolated it can readily be inserted in a straightforward manner into a plasmid for cloning. The procedure is diagrammed by Williams (10) and Adams et al. (11). The essential steps are as follows:

The plasmid such as pB322 containing antibiotic resistances (e.g., Amp^r and

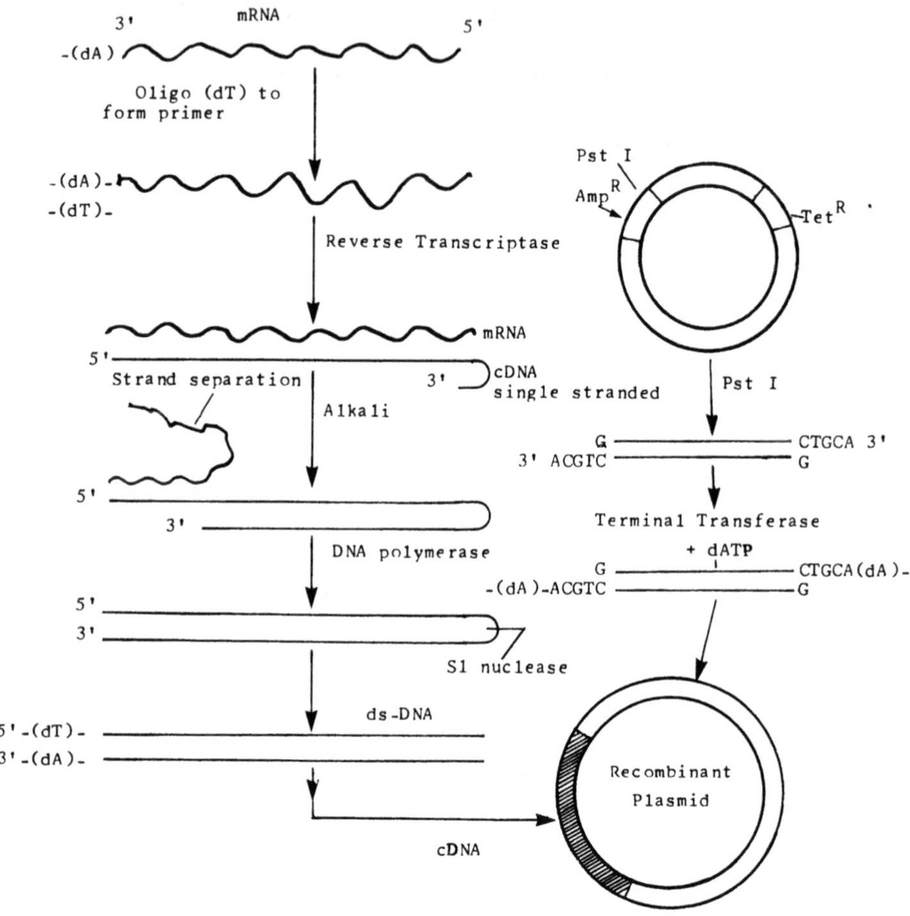

FIGURE 9 Preparation of eukaryotic mRNA for insertion into plasmid. Eukaryotic mRNA is converted to cDNA by reverse transcription and tailored to fit into a plasmid.

Tet^r) is cleaved in a conventional way with *Pst*I (reagent mixture for example: 10 m*M* Trischloride, pH 7.4, 50 m*M* NaCl, 6 m*M* $MgCl_2$, 6 m*M* β-mercapto-ethanol, 100 μg/ml gelatin or bovine serum albumin). Terminal transferase will add poly tails using dATP.

Annealing

The homopolymers were selected so that the recombinant sequence retains the *Pst*I target sites; this means that Tet^r has been retained but the structure has acquired Amp^s, ampicillin sensitivity, because the original *Pst*I cleavage cut the Amp^r. Final sealing of nicks with DNA ligase yields reformed plasmid ready for cloning.

Use of Cosmid Vector for Cloning

The gene map illustrated in Figure 10 shows that the nonessential region is available for the insertion of a suitable DNA fragment. The plasmid structure containing the *cos* site can be opened with a restriction endonuclease and the foreign DNA is ligated to it; this DNA is cleaved to fit opened plasmid.

The products of ligation represent a variety of entities, an essential linear unit of limited kilobase length is the desired sequence, so that it is small enough for packaging into a phage particle. Some other structures are also formed, mostly circular entities, for instance recircularized cosmid, but their constitution does not lead to clonable units.

The recombinant structure, the *cos*-carrying linear DNA fragment is readily introduced into a phage; it behaves like a phage DNA and circularizes. After phage adsorption on a host and DNA injection, the cloning process proceeds as usual. Figure 10 shows the sequential steps.

In a brief note, written for scientists in general, Yanchinski (12) characterizes cosmids in a unique style to wit:

> The cosmid combines the advantages of the plasmid's small size with the phage's ruthless drive to penetrate the bacterial cell wall.

CLONING IN MAMMALIAN CELLS

The simian virus 40 (SV40) was briefly described in Chapter 6. As an example of *cloning in mammalian cells* the construction of a hybrid consisting of SV40 and λ DNA is illustrated in Figure 11. The study, conducted by Goff and Berg (13), was discussed in some detail by Old and Primrose (9). The wild-type SV40, cannot be used directly for insertion of λ DNA, as such a combination would be too large to be packaged into a viral particle. Therefore, some of the SV40 viral DNA is excised, and the essential remainder is reacted with a desired fragment

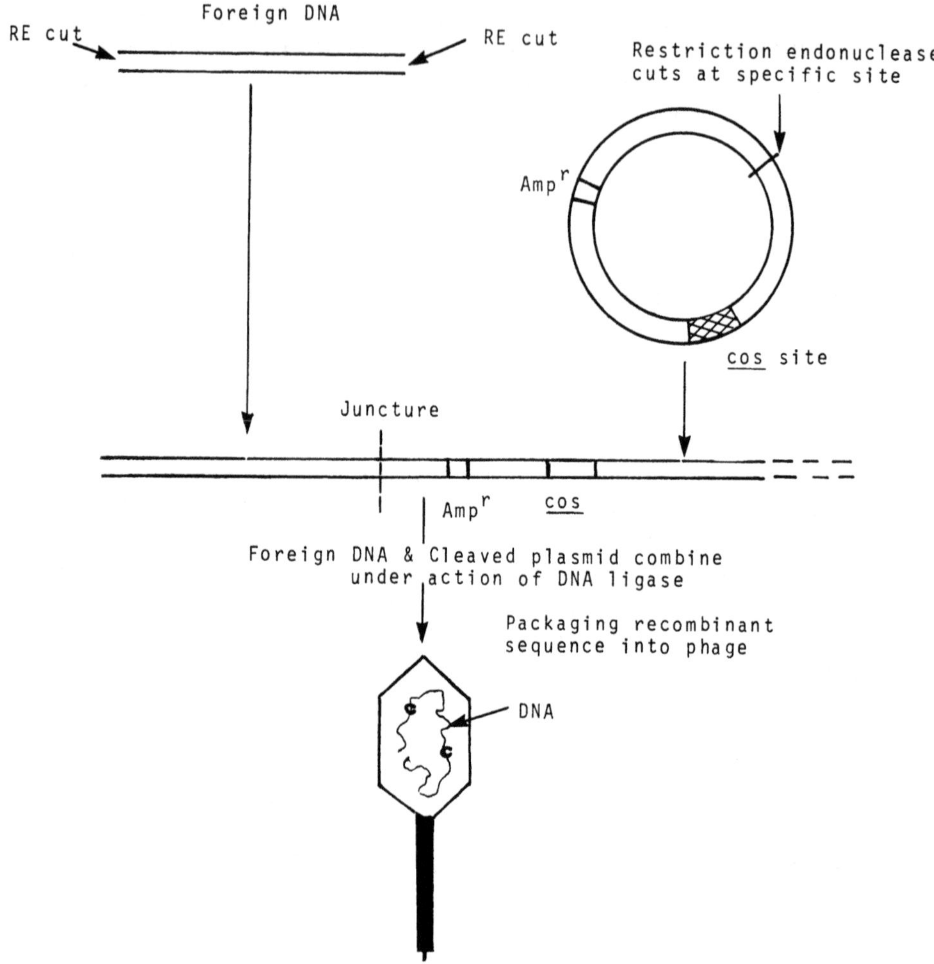

FIGURE 10 Cosmid vector for cloning. Linearizing cos vector and combining with selected foreign DNA, followed by packaging in a virus, and subsequent injection of viral DNA into a receptive cloning vehicle.

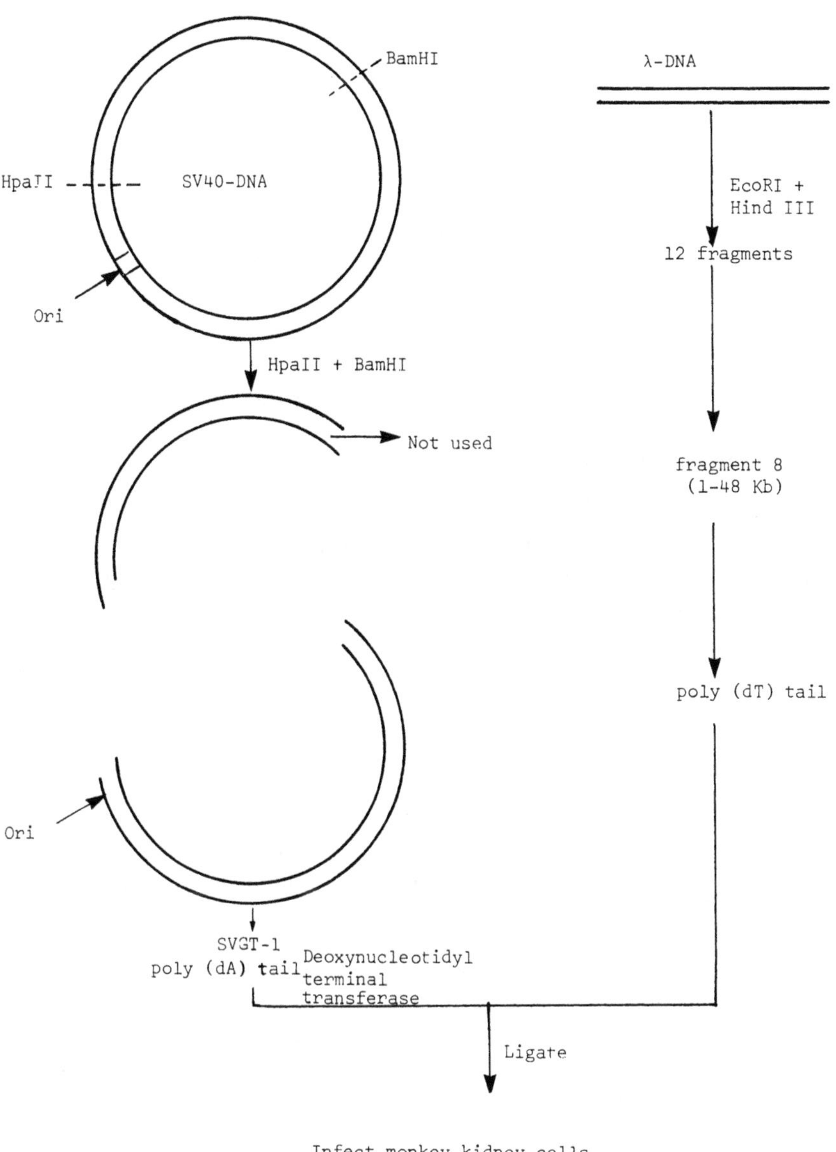

FIGURE 11 Construction of SV40-λ DNA hybrid. SV40 circular DNA is trimmed and recombined with linear λ DNA. Diagram was constructed from information contained in References 13 to 17.

from λ DNA. As the procedure is a very good example of recombinant techniques it is worthwhile to excerpt some of the details from the publication.

Modification of SV40

The circular SV40 structure was cleaved serially with *Bam*I and *Hpa*II to excise the so-called late section (see Fig. 11) and to use the early section in the hybridizing scheme. An essential feature was the retention of the origin of replication, the *Ori* in the early section. This fragment was labeled SVGT-1. The structure was then modified to remove about 50 nucleotides from the 5′ end of each strand; an excess of λ exonuclease (λ E) accomplished this excission. Further incubation with deoxynucleotidyl terminal transferase (dtt) and [^{3}H] dATP added about 300 deoxyadenylate residues per exposed 3′ end (i.e., poly (dA) tail).

Preparation of Phage λ DNA Insert

The phage λ DNA used contained five *Eco*RI cleavage sites and six *Hind*III sites. Upon digestion with both enzymes, 12 fragments were obtained which ranged from 0.6 to 23 kilobase pairs (kbp) in length. Separation of fragments was achieved by electrophoresis on agarose gel. Fragment number 8 at 1.48 kbp was considered most suitable; it also contained desirable genes, transcriptional promoters, and *Ori*. This fragment was then treated with λE followed by reaction with dtt to add poly (dT) tails.

Formation of Hybrid

The poly (dA)-terminated λSVGT and the poly (dT)-ended λ DNA were mixed, heated to 68°C for 10 min, cooled to 47°C, and kept there for 2 h. Annealing took place in this phase, and further cooling to room temperature in about 5 h, completed the process.

After this ligation the combination was tested in monkey cells to confirm the existence of the expected hybrid. The SVGT- λ hybrid DNA did replicate in the cultured monkey cells when tsA mutant helper DNA was used. However, the λ DNA sequences were not transcribed. In spite of this failure, the procedure was valuable in proving its potential usefulness. The original publication contains a discussion of possible reasons for the lack of transcription. A similar recombinant construction attempt was reported by Southern, et al. (18), who inserted a cDNA segment, which contained coding for rabbit β-globin, into a SV40 structure.

COMMERCIALIZATION

As an example of a neatly devised sequence of biochemical reactions a diagram from a commercial bulletin is very instructive (19). Obviously, the use of such an

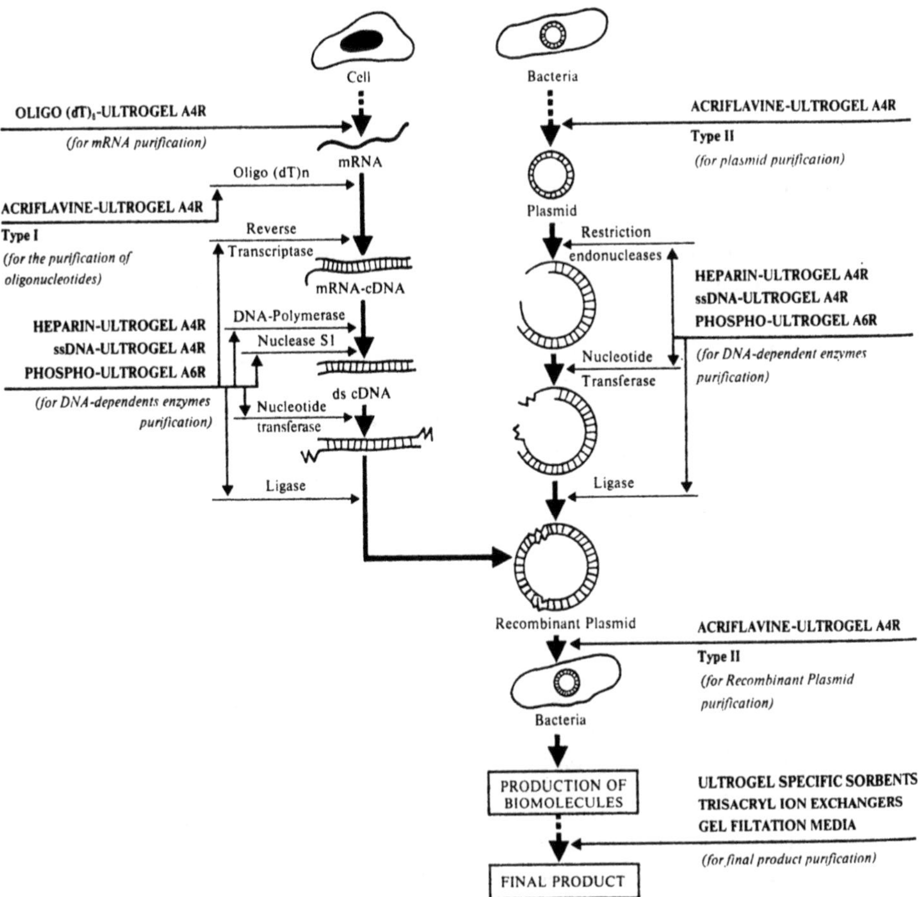

FIGURE 12 Commercial recombinant strategy. Outline of classical recombinant procedures involving intersequential gel separations. Listing of specific commercial gel types is instructive (see text). (Courtesy Reactifs IBF, Villeneuve-la-Garenne, France). (Ref. 19).

illustration does not constitute a recommendation of specific manufactured products. However, when an illustration is particularly informative, its source should not mitigate against using it for instructional purposes. The intelligent reader is fully aware of the availability of reagents from a variety of sources (see Fig. 12).

As such, the diagram presents the bread-and-butter sequence of events. Cleaving of ingredients, separation of fractions on a gel, identifying and isolating desired sequences, excission from the gel and elution. Also reverse transcription, poly-tailing, and recombinant ligation. The diagram thus demonstrates the use of appropriate reactants and throws light on the manifold commercial availability of reagents. Additionally, the illustration points up the numerous intersequential steps involving gel separations with gel types suited to particular procedures.

The company, Reactifs IBF (19), also makes available a very instructive "Practical Guide for use in affinity chromatography techniques." The bulletin is in English, comprising 157 pages and lists 474 references, so it is a worthwhile addition to one's library.

REFERENCES

1. Grossman, L, Moldave, K. (1980). *Methods in Enzymology. 65*, Part I, Acad. Press, New York, p. 511.
2. Pelczar, MJ Jr, Reid, RG, Chen, ECS. (1977). *Microbiology*, 4th Ed. McGraw-Hill, New York.
3. Morris, ChE. *Food Engineering* 53 (5): 57–69 (1981).
4. Dally, EL *et al.* Recombinant DNA technology-Food for thought. *Food Technology* 35 (7): 26–33 (1981).
5. Biosearch, San Rafael, CA 94901; Linkers and Adapters Bulletin.
6. Bernard, HU, Helinski, DR. (1980). Bacterial Plasmid Cloning Vehicles. In *Genetic Engineering Principles and Methods*, Vol. 2, pp. 133–167, Eds. Setlow, JK. Hollaender, A (Eds), Plenum Press, New York.
7. Roberts, RJ (1980). Directory of Restriction Enzymes. In *Methods in Enzymology 65*. Academic Press, New York, Part I, pp. 1–15.
8. Cohen, SN, Construction of Biologically Functional Bacterial Plasmids *in Vitro*. *Proc. Natl. Acad. Sci.*, (*USA*) 70 (11):3240–3244 (Nov. 1973).
9. Old, RW Primrose, SB. (1980). *Principles of Gene Manipulation, Studies in Microbiology*, Vol. 2. University of California Press, Berkeley, CA. Blackwell Scientific.
10. Williams, JS (1981). The preparation and screening of a cDNA clone bank. In *Genetic Engineering* I, Williams, R (Ed). Acadmic Press, New York.
11. Adams, RLP et al. (1981). *The Biochemistry of the Nucleic Acids*, 9th Ed. Chapman and Hall, New York.
12. Yanchinski, S. Cosmids. *New Sci. 95* (1312): 23 (1982).
13. Goff, SP, Berg, P. Construction of hybrid viruses containing SV40 and λ

phage DNA segments and their propagation in cultured monkey cells. *Cell* 9 (Part 2): 695–705 (1976).

14. Collins, J, Hohn, B. "Cosmids: A type of plasmid gene-cloning vector that is packageable *in vitro* in bacteriophage λ heads. *Proc. Natl. Acad. Sci.* (*USA*) 75 (9): 4242–4246 (1978).
15. Hohn, B, Murray, K. Packaging recombinant DNA molecules into bacteriophage particles *in vitro. Proc. Natl. Acad. Sci.* (USA) 74 (8):3259–3263 (1977).
16. Collins, J, Brüning, HJ. Plasmids Useable as Gene-cloning Vectors in an *in vitro* Packaging by Coliphage λ: Cosmids. *Gene 4* (2): 85–107 (1978).
17. Gottesman, ME, Yarmolinsky, MB. (1968). The Integration and excission of the Bacteriophage Lambda Genome. *Cold Spring Harbor Symposia on Quantitative Biology*, Vol. XXXIII, pp. 735–747.
18. Southern, PJ et al. Construction and characterization of SV40 recombinants with β-Globin cDNA substitutions in their early region. *J. Mol. Appl. Genetics* 1 (3): 177–190 (1981).
19. Société Chimique Pointet-Girard, Reactifs IBF, 35 Av. Jean Jaurès, 92390 Villeneuve-La-Garenne, France (Advertisement in *Nature*); Also Baschetti, E et al. *Fed. Eur. Biochem. Socs.* 139 (2):193–196 (1982); Egly, JM et al. *J. Chromat.* 243: 301-306 (1982); Egly, JM et al. (1982). *Affinity Chromatography* Elsevier, New York; also Bulletin on Affinity Chromatography, 2nd Ed. (1983).

8
Nucleotide Sequencing and Hybridizing

Nucleotide sequencing is the analytical chemistry of genetic engineering in general and recombinant techniques in particular. The procedure uses the well-known method of electrophoresis where organic molecules of different size will travel at different rates in a gel matrix under the influence of an electric potential (direct current); larger molecules move more slowly because of their interaction with the gel structure. Thus, DNA chains of varying nucleotide content (complexes of P–S–B)* which have been labeled by ^{32}P insertion at 5′ end will spread on the gel into discrete sequential position. The location of the respective nucleotides is visualized on the gel by staining with ethidium bromide (EtBr) or a fluorescent dye. The EtBr or dye fluoresce under ultraviolet light and a photographic record can be established. Also, when only small amounts of DNA are present, subsequent exposure to an x-ray film gives an autoradiograph of ^{32}P location which then can be read to ascertain the order of bases.

Before electrophoretic gel separation can be performed, a number of fairly involved preparative steps must be undertaken. All sequencing methods presently in use require the production of a series of DNA fragments. Only when knowledge about and utilization of restriction endonucleases matured was it possible to create orderly procedures for making definitive nucleotide fragments from the normally very long DNA molecules. The process of fragment analysis is called "restriction mapping." A particular feature of sequencing methods is the production of fragments which have one common end and the other end in varying positions. Operationally, four types of series of fragments are created in such a way that each group is terminated by one of the four bases: A,C,G,T.

---

*P–S–B is phosphorus-sugar-base.

THE SEQUENCING GEL

Compositions containing 8–10% polyacrylamide or ~ 0.8% agarose are commonly used to prepare gel slabs for sequencing. A number of such gel formers are available commercially. The solution is usually adjusted to pH 8.3, to make sure that all nucleotides move to the anode. To prevent the formation of intermolecular or intramolecular double-stranded regions, a denaturant (i.e., 7 *M* urea) is added to the gel formation.

The gel thickness is adjusted by the operator to suit the expected conditions. A thin gel slab 0.4 to 0.5 mm thick will give less radioactive scattering and thus sharper bands on the autoradiograph. Also, with thin gels a higher voltage can be used leading to a faster operation. A thick slab would be about 1.5 mm deep. The time allowed for electrophoresis is determined by what one wants to have on the final gel sequence. If a short time is used, the smaller fragments will be captured; with a longer operating period the larger fragments are fixed, while a number of smaller fragments "run off."

SEQUENCING METHODS

A detailed treatment of all of the typically used methods would be lengthy and redundant. Thus, only brief remarks are in order for two of the more frequently used procedures and a more thorough coverage for the method which appears to be the most widely used one. As with many other procedure involving DNA manipulation, the development of these methods was made possible by the availability of a large variety of restriction endonucleases (enzymes), that had not been accessible up to then. Using one or more such enzymes, almost any section of DNA can be fractured into lengths of 100 to 200 or more nucleotide pairs, lengths which are needed in the above sequencing methods.

DNA Sequencing

A comprehensive review of sequencing, that is, DNA sequence analysis, by Air (1) discusses in considerable detail two preferred methods. They are, the Sanger and Coulson "plus and minus" method (2) and that of Maxam and Gilbert (3). The basic difference between these methods is that Sanger and Coulson use a so-called copying method with a staggered primed DNA synthesis to obtain a mixture of proper length fragments which are then reacted with nucleoside phosphates, while Maxam and Gilbert employ partial "chemical degradation" to accomplish DNA cleavage. Both methods rely on ^{32}P labeling needed for later autoradiography. Another method of interest is the Southern blot technique (4, 5) for hybridizing DNA strands to determine if there is homology.

RNA Sequencing

A procedure for sequencing RNA has been described by Kramer and Mills (6). By including 3′-deoxy analogs of ribonucleoside triphosphates during RNA synthesis, it is possible to create a strand ending with a 3′-deoxyuridine which has been incorporated in place of uridine. However, the most common procedure at present is to make a complementary copy of the RNA, a cDNA, by the use of reverse transcriptase and then analyze by DNA sequencing.

SPECIFIC METHODS

Sanger and Coulson "Plus and Minus" Method

The method involves two stages of DNA synthesis and in principle is a copying method based on the formation of a hybrid DNA on a template. The starting material is a *single-stranded DNA template*, the sequence to be analyzed. A complementary sequence is then annealed as a primer. The primer can be an oligonucleotide which has been prepared synthetically or a double-stranded fragment obtained by cleaving a DNA molecule with a restriction enzyme. A condition is that only the fragment strand that is complementary to the template will anneal. The four deoxynucloside triphosphates (one of which is labeled with α-^{32}P) are added together with *Escherichia coli* DNA polymerase I. DNA synthesis, the copy strand, takes place in a random manner. As polymerization progresses small samples are taken, so that upon termination a collection of different length fragments is available. All samples are combined and the synthesized mixture is divided into 8 equal parts.

Minus System. The *four* samples are treated with *three* nucleoside triphosphates each, which will add on to the fragments, but addition will stop when the missing triphosphate is specified by the template. The result is that each of the four sample mixtures has ends so that the next base would be A, C, G, or T. For instance the "-G" mix contains a set of oligonucleotides of different lengths, and one of the molecules would be such that the next residue would be a G. The restriction fragment primer is then removed by cleavage and the template is dissociated by heating in formamide. Running the four samples on an electrophoretic gel and making an autoradiograph yields the typical sequence x-ray picture.

Plus System. Each of the four samples is mixed with one deoxyribonucleoside triphosphate. Treating with DNA polymerase obtained from phage T4-infected *E. coli* results in double-strand degradation starting at the 3′ end. If a single nucleoside triphosphate is present, the enzyme action stops. Each sample will contain only those molecules in which the respective base is the 3′ nucleotide. Sequences are then read as in the minus method.

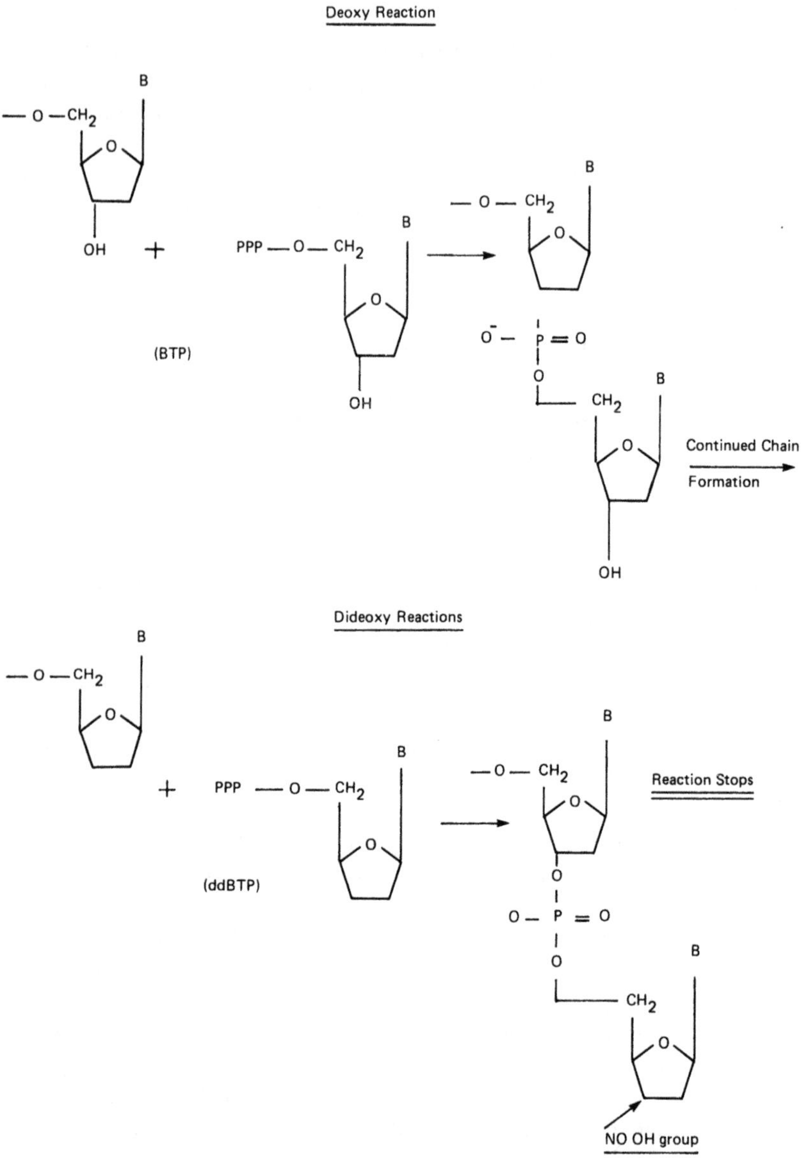

Deoxy Reaction

(BTP)

Continued Chain Formation

Dideoxy Reactions

(ddBTP)

Reaction Stops

NO OH group

FIGURE 1 (a) Comparison of deoxy and dideoxy reaction. Upper part is regular deoxy compound addition. Lower part involves a dideoxy-compound, leading to chain termination.

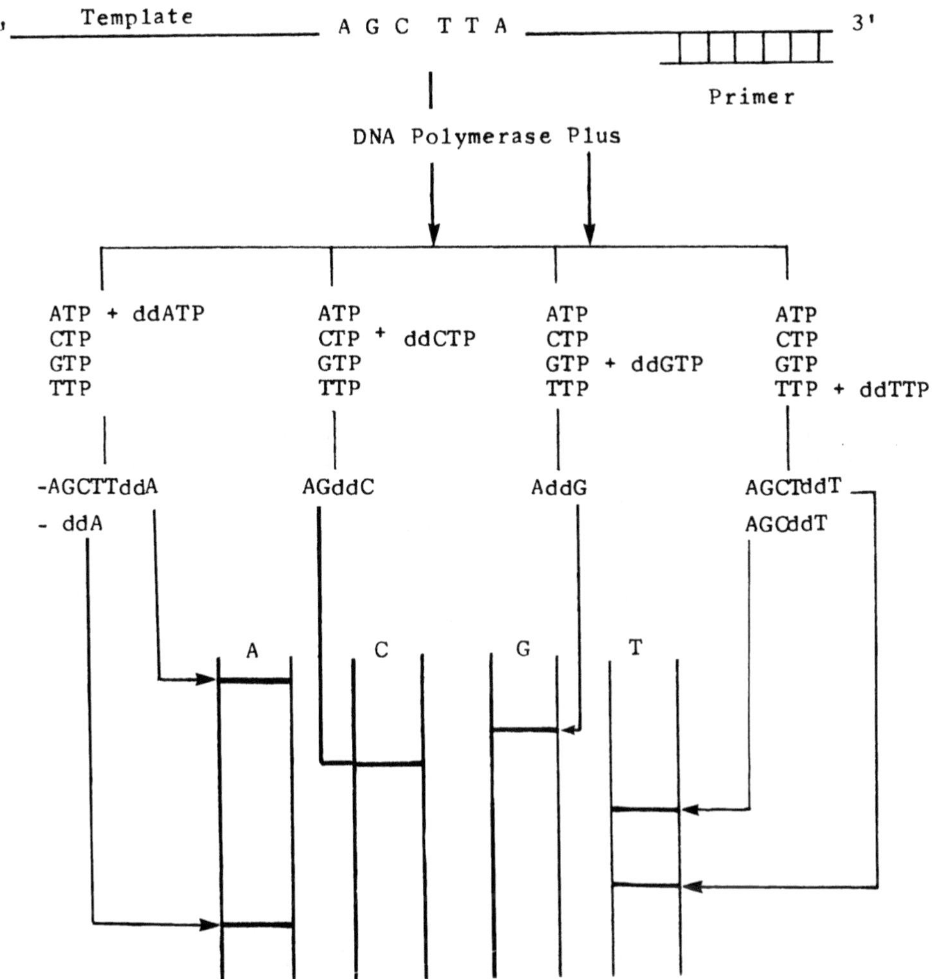

(b) Schematic of Sanger dideoxy sequencing. The number of chain fragments ending with dd-base shows up in the respective lanes of the bases; e.g., first lane: There are two fragments ending in ddA and so there are two bands in the A column. Respective locations of the dd-bands are determined by length of fragment: the shortest being near the bottom of the gel plate (Ref. 7).

The plus and minus was one of the original accomplishments which contributed to the reputation of the Sanger group. The method is used sparingly, as it was superceded by the "dideoxy" method. As such, the P&M method, however, is certainly of considerable historical interest and it contains important procedural developments which are common to most of the sequencing methods in use today.

Sanger, Dideoxy Method

The dideoxy method (2,7) uses the chain-terminating action of dideoxynucleoside triphosphates. They are essentially like the normal deoxynucleoside triphosphates, but the 3′-OH group is absent. This means that they will readily add to a growing DNA chain but then there is no 3′-OH group for further reaction and the chain growth stops. The difference in the reactions is clearly illustrated in Figure 1a, which traces the normal deoxy reaction through esterification at the phosphate site. Here, the 3′-OH plays an essential role. The dideoxy reaction is illustrated in the lower part, where the absence of the OH group leads to chain termination.

Most often, the dideoxy method is used by first cloning the single-stranded template DNA in M13 phage, with the addition of primers at the flanks of the DNA insert. The schematic of the dideoxy procedure from template DNA through four dd-nucleoside triphosphate-laced samples to electrophoresis gel separation is shown in Figure 1b. (*Note.* The sequence frequency in the lanes corresponds to the number of dd-bases formed. The structure sequence appearing on the gel is from the complementary fragments. Thus, the template DNA sequence will be read by base pairing.)

Southern Blot Technique: Hybridizing

Restriction fragments of DNA resulting from endonuclease digestion and gel electrophoresis can be denatured by treating with alkali and transferring to strips of cellulose nitrate filter paper. When radioactive RNA is applied to the filter paper, complementary DNA sequences will combine with the RNA to form a hybrid structure. After washing the filter with a citrate buffer solution, the radioactivity can be visualized by autoradiography or fluorescence photography. The readings permit selection of DNA fragments that carry the complementary sequences of the applied known RNA composition. The technique was developed by Southern and was published in 1973 (4). The original article contains detailed directions for conducting the procedure.

While Southern's method was frequently used as an adjunct to other sequencing procedures, it is now recognized that its principle is a very valuable concept which can be applied to the sensitive detection of any biologically important macromolecule. A concise discussion of such application is offered in a company bulletin by BIO RAD Laboratories (5). There, Southern, Northern, Western, and electro blot schemes are delineated in a highly condensed form with appropriate literature references. The bulletin is readily attainable and must be recommended to all engaged in sequencing problems.

CHEMICAL SEQUENCING METHOD

The Maxam and Gilbert method was originally published in 1977. More recently (3) these authors presented an updated version. Their introductory statement concisely describes the procedure as follows:

> In the chemical DNA sequencing method, one end-labels the DNA, partially cleaves it at each of the four bases in four reactions, orders the products by size on a slab gel, and then reads the sequence from an autoradiogram by simply noting which base-specific agent cleaved at each successive nucleotide along the strand. This technique will sequence DNA made in and purified from cells. No enzymatic copying *in vitro* is required, and either single- or double-stranded DNA can be sequenced. . . .

The Maxam and Gilbert method is so widely used, that a somewhat more detailed coverage is indicated. As an example, the identification of base sequences in a *plasmid DNA* will be most instructive.

1. The cloned plasmid DNA is double-stranded and circular. (It often contains genes which impart penicillin and/or tetracylcine resistance; in general an antibiotic-resistant genome.)
2. The circular DNA is cleaved at a specific site by means of a suitable restriction enzyme. This results in linearized DNA strands. The aim is to generate DNA fragments with so-called "unique ends."
3. The fragments from Step 2 are separated on a gel, acrylamide, or agarose, by electrophoresis.
4. The desired subfragments are isolated by excision of the gel portion and elution of the DNA.
5. The 5′ ends of the subfragments are labeled with ^{32}P in solution, using DNA polymerase 1 (*E. coli*) at 3′end $+a^{32}P$ and kinase at 5′end $+\gamma^{32}P$ (6).
6. For identification, single-labelled ends are needed. The double-stranded DNA, marked at both the 3′, and 5′ positions must be converted to single ends by:
    a. Strand separation: denature DNA with NaOH and isolate single strands on a neutral gel (electrophoresis).

   b. Secondary digestion with a restriction enzyme that cleaves between ends. This leaves single ends.
7. Stepwise chemical treatment proceeds as follows:
   a. The chemical agent modifies one or more of the four DNA bases, by substituting in a purine or pyrimidine ring.
   b. The second reaction removes this modified base from its sugar.
   c. A third reaction eliminates both phosphates from that sugar to break the DNA.
8. a. The ^{32}P-labeled subfragments are divided into 4 fractions for A, G, C+T, and T identification.
   b. Each fraction is subjected to a chemical treatment to destabilize the phosphodiester linkage. For: *A* use NaOH + EDTA at 90°C (temperature may vary with time). *G*, use dimethyl sulfate at 20°C. *C + T*, use hydrazinne at 20°C, *T*, use hydrazine + NaCl at 20°C. The objective is to modify ~ 1% of each particular nucleotide in the fragment.
   c. Chemical cleavage of the modified bases (located in nucleotides) is then accomplished by heating the respective samples in 1 *M* piperidine solution at 90°C for 30 min.

The example of the modification of guanine with dimethyl sulfate is worthy of elaboration. The sequence is as follows:

Treatment of the guanine nucleotide with dimethylsulfate results in methylation of the 7-N position. The alkali addition causes the opening of the imidazole ring (the five-membered ring) between the 8 and 9 positions. The piperidine reaction leads to the separation of the remaining guanine structure and formation of a piperidine adduct with the opened ribose sugar and two phosphate radicals. The net result is that the guanine structure and the ribose structure have been removed together with two phosphate links resulting in the breakage of the DNA polymer at that position. The procedure is illustrated, without details of intermediate steps, in Figure 2.

Continuing with the example of guanine removal, a DNA sequence will break at every G location and then yield a number of sequential fragments of varying length. Each fragment will end with the base immediately preceding the G bases. Schematically, the sequential steps are presented in Figure 3. The figure also shows the base sequences resulting from the multiple cleavages at the G locations. Both illustrations are patterned after Maxam and Gilbert's original diagram with permission by Gilbert.

It should be noted that an artist's conception is contained in the Maxam and Gilbert article (3). It might be superfluous to mention that their illustration shows two five-membered rings at the G positions, which could be misinterpreted by the nonbiochemist. Thus, the upper five-membered ring (Fig. 1, page 500 of the reference) represents the deoxybase sugar, and the adjacent lower

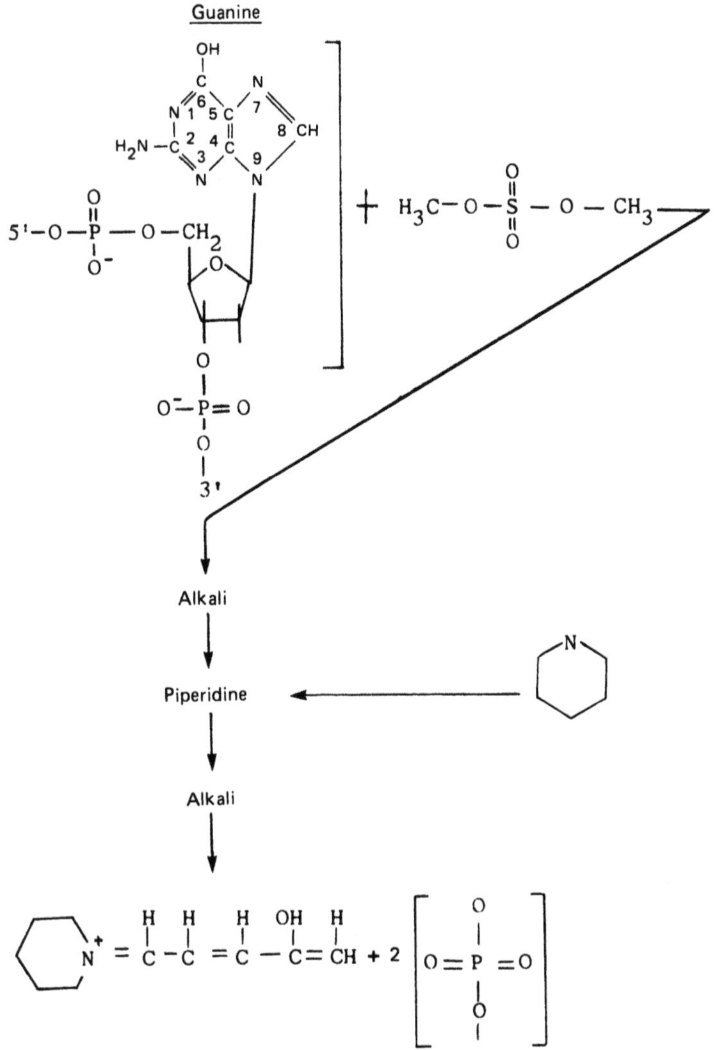

FIGURE 2 Chemical break at guanine. Dimethyl sulfate and piperidine break the DNA at guanine. DMS methylates the N-7 atom.

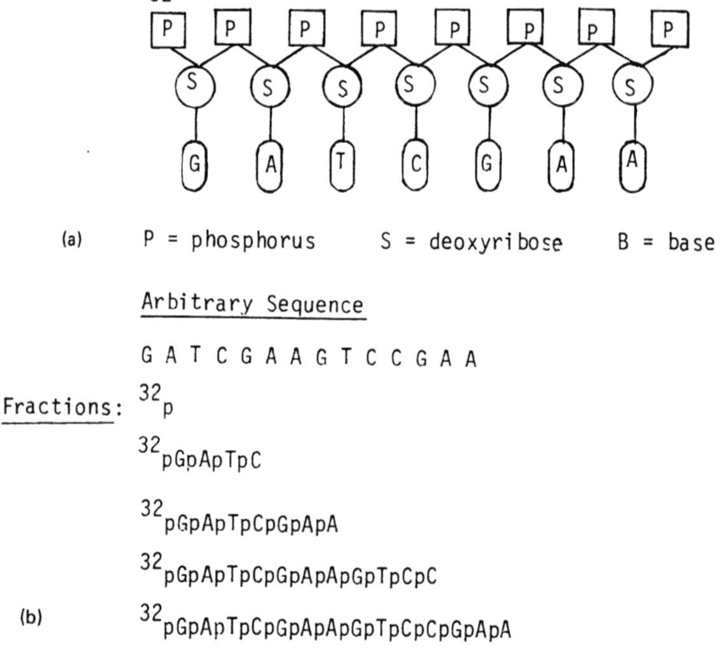

FIGURE 3 Base-specific cleavage of DNA (a) Modified bases are progressively cleaved. Methylated G. (b) Nested set of end-labeled fragments.

five-membered ring is the imidazole portion of the guanine molecule. The p notation signifies unlabeled P atoms. The cleavage leaves a phosphate at one end and an OH^- radical at the other end of the cleavage site.

After chemical treatment each of the four portions is treated several times with ethanol to precipitate the DNA strands and centrifuged to obtain a DNA pellet. Finally, the four sets of cleavage product are dissolved in a formamide -NaOH solution for electrophoresis (a marker dye is usually added to visualize the progress). These solutions are heat-denatured and loaded into separate wells onto a polyacrylamide (or agarose) sequencing gel.

On the gel, with appropriate voltage (lower voltage for thicker gel, for instance, 800 V for 1.5 mm thickness and up to 4000 V for thin gel) and use of electrophoresis buffer, the larger fragments move slower than the smaller ones because of slightly greater noncovalent interaction with the gel matrix. (*Note.* Voltage must be such that the *gel* will *not heat* up.)

When the electrophoresis is completed (perhaps 30 h) the slab gel, still in place on one of the glass plates is covered with a plastic sheet, for example Saran, and the x-ray film is laid on the plastic (or the gel assemply onto the x-ray film), weighted down and kept at -70°C for 1 to 5 days. The ^{32}P labeling will give dark streaks which then can be interpreted as follows:

DNA Sequence Band	A	G	C+T	C
		Band Location		
G		———		
		(1)		(2)
C			———	———(2)
A	———			
	(3)			
T			———(4)	
T			———(5)	
G		———		
		(6)		
A	———			
	(7)			
C			———	———(8)
C			———	———(9)
G		———		
		(10)		
A	———			
	(11)			

The sequence read from this diagram is then: *GCATTGACCGA*.
Example of an "A" Lane Run
The nucleotide sequence is:
- -$(N)_y$ - A-G-A-C-T-G-A- -(x* labeled end)
Fractions obtained are:
- -(N)Y - A + G-A-C-T-G-A-(x*)
- -$(N)_y$ - A-G-A + C-T-G-A-(x*)
- -(N)Y - A-G-A-C-T-G - A+(x*)
The fractions ending in A will appear in the *A* column of the gel slab.

Preparation and Installation of Gel Slab

The gel solution is degassed; the degree of degassing will affect the polymerization time. The glass plates 12 inches wide by 36 inches long are clamped together with a spacing of 0.4 mm. The bottom and sides are clamped. The assembly is then supported in an inclined position and the gell solution is poured into the top. Filling of the cavity occurs by gravity and capillary action. The top well (for receiving the electrophoresis solution) is attached. The assemply is kept at room temperature until polymerization is complete (a test specimen in a test tube is observed). The setting time may vary from 12 min to 2 h. The polymerized gel assembly is placed in a vertical frame and connected to the voltage regulator.

Example of Unusual DNA Sequence

Sequencing operations most of the time are fairly straightforward. Occasionally difficulties arise because the equipment malfunctions and also when a streaking of the gel plates occurs because the molecular separation is not clean enough, and the bases overlap in the gel. Such situations can usually be corrected without too much trouble.

Sometimes, however, some "artifacts" are encountered which are real, but defy explanation. The term artifact is used by the biochemist to denote an unusual phenomenon, which could or could not be significant.

Separation Process

The so-called free flow electrophoresis procedure, a well-known separation process has been developed to a completely automated stage. It is a procedure which may be of considerable help as an adjunct in recombinant research.

Migration of molecules and also cellular material on the basis of differences in electrophoretic mobility will result in effective separations.

Some of the applications are:

- Micro- and macromolecules
  - Proteins
  - Mucopolysaccharides
  - Enzymes
- Cell organelles
  - Phages (e.g., selection of mutants)
  - Bacteria
  - Parasites

The experimental set up is very similar to that of gel electrophoresis but the separation chamber is an open-flow channel of about 0.5 mm in depth. Buffered solution flows downward and perpendicular to an electrical field. A sample inlet just below the buffer entry is used for sample injection. Separation of substances then is in an angular direction. The process is continuous, so injection and sample collection proceed with time as long as one wants to conduct separations.

A particularly effective version of the equipment is manufactured by MSE Scientific Instruments (8). The automatized unit is illustrated in Figure 4 and the separation scheme is diagrammed in Figure 5. Operating time of 6 to 12 h is about maximum and sample throughput per hour in the range of 30–250 mg protein and 30-500 x 10^8 cells. Scanning of the operation is conducted by means of an optical-electronic combination, permitting continuous recording as a pherogram. An excellent bulletin is available which describes the ElphorVaP unit in considerable detail. This process should be considered as an alternative choice, not as a replacement of well-established conventional methods.

Potential Developments

From the viewpoint of the chemical analyst, the gel electrophoretic system to separate DNA mixtures is a somewhat fuzzy analysis. Nevertheless, it is the only workable procedure in existence, as exemplified by the Sanger and the Maxam-Gilbert sequencing methods.

A welcome review and analysis of potential improvements was published recently by Martin and Davies (9). Most of this treatise is concerned with the automation of the existing procedures and other possible analytical methods are mentioned only briefly. A quotation from the report well describes the present situation as "Extensive fundamental investigations are required in separation science to develop a continuous system suitable for nucleic acid fractionation to produce unambiguous band data suitable for image abstraction. Improvement to band imaging techniques and data interpretation software are required."

FIGURE 4 Free flow electrophoresis unit. Elphor VaP unit manufactured by MSE Scientific Instruments (Ref. 8).

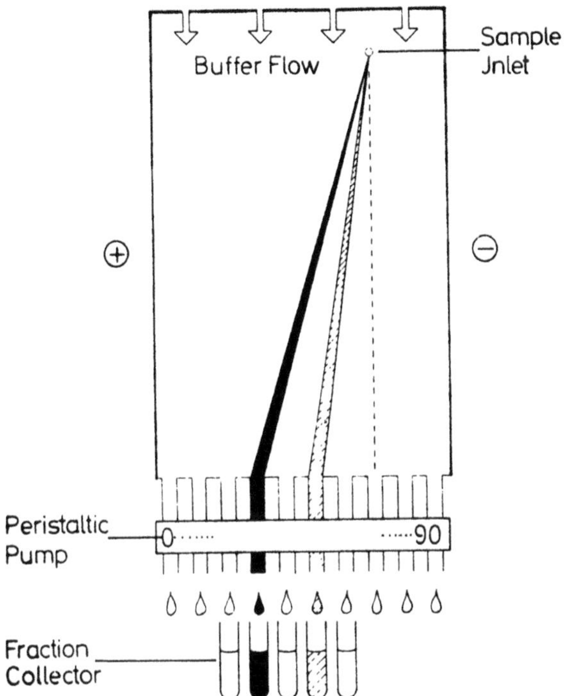

FIGURE 5 Flow scheme in FFE. Angular displacement is a function of size and due to electric field. (Ref. 8).

REFERENCES

1. Air, GM. Rapid DNA sequence analysis. *Crit. Rev. Biochem.* 6(1): 1–33 (1979).
2. Sanger, F, et al. DNA sequencing with chain-terminating inhibitors. *Proc. Natl. Acad. Sci.* (*USA*) 74(12):5463-5467 (1977); also: *J. Mol. Biol.* 94 (2): 441–448 (1975).
3. Maxam, AM, Gilbert, W, (1986). Sequencing end-labeled DNA with base-specific chemical cleavages. In *Methods of Enzymology*, Vol. 65, Part I, pp. 499-560.
4. Southern, EM, Detection of specific sequences among DNA fragments separated by gel electrophoresis. *J. Mol. Biol.* 98(3):503–517 (1973).
5. *Southern, Western and Electro-Blotting.* BIO–RAD Laboratories Bulletin 1980.
6. Kramer, FR, Mills, DR. RNA sequencing with radioactive chain-terminating ribonucleotides. *Proc. Natl. Acad. Sci.* (*USA*) 75 (11):5334–5338 (1978).

7. Sanger, F. Determination of nucleotide sequences in DNA. *Science* 214 (4528):1205–1210 (1981).
8. MSE Scientific Instruments, Manor Royal, Crawley, West Sussex RH10 2 QQ, England.
9. Martin, WJ, Davies, RW. Automated DNA Sequencing: Progress and prospects. *Biotechnology* 4(10):890–895 (1986).

9
Plant Genetics

Attempts to create genetically superior plant species predate the age of genetic engineering by almost two centuries. Lacking the knowledge and ability to influence genes directly, the studies had to be carried out by cross-breeding. Nevertheless, some notable successes have been achieved, such as Luther Burbank's fruit miracles, Barbara McClintock's pioneer studies with maize (1), Norman Borlaug's green revolution in high yield wheat and rice, extensive cross pollenation for seed corns, a host of novelty developments in horticulture, refining the tissue culture process to create what amounts to cloning by using plant growth hormones with proper plant nutrients, creation of somatic cell fusion (2), as well as a multitude of other inventive improvements too numerous to list.

Scientists engaged in plant genetics point out that there are three basic problems in the genetic engineering manipulations (3). They are the introduction of the genetic material, the expression of the introduced material in the host cell, and the regeneration of the whole plant, where the last step is likely to be the most difficult task.

GENE TRANSFER PROCEDURES

Non-recombinant

The significance of molecular biology applied to plant science was reviewed by Fox (4). There are as yet no definitive results from recombinant techniques, although the potential for intensive applications exists. Methods other than direct genetic intervention on the DNA level, but still of a nature to be considered genetic engineering, have been used with variable success.

Cell Fusion Technique

An enormous amount of research is underway to improve or modify plant species by means of cell fusion. Although this procedure is not a recombinant technique as such, it qualifies as a genetic engineering task.

A fascinating publication by Blair (5) highlights the manipulation of *protoplast fusion* in plant tissue culture and describes a number of successful hybrid creations. A prominent activity is that of Cocking and co-workers (2); also a volume on protoplast research (6).

Protoplasts are created by dissolving the plant cell wall with an enzyme, a lysozyme. The removal of the cell wall leaves intact bodies, the protoplasts. When protoplasts from two different plants are brought together, fusion of some of the cells will take place. This phenomenon is called protoplast fusion or somatic cell fusion. A suitable liquid medium for cell suspensions is polyethylene glycol (PEG). Experience shows that as much as 10% of the two types of nonrelated cells will fuse. The fused protoplasts are then introduced into a solution containing appropriate amounts of cytokimine and auxin to promote regeneration of the cell wall.

Thus protoplast fusion can result in the introduction of disease-resistant traits, increased production of wanted plant constituents, hybrid foods, and climate-tolerating plant species. Specific examples described by Blair are: production of tobacco hybrids, lettuce varieties, a potato-tomato hybrid, asparagus size regulation, pyrethrum flowers with greatly increased productivity, dwarf fruit manipulation, substantial advances in eucalyptus tree culture, improved oil production from the gopher plant, upgrading of palm trees to increase palm oil yields, establishing grape vines that are resistant to Pierce's disease, insuring survival of threatened plant species, and lysine-rich grain amaranth cultivation.

The specific application of protoplast fusion in the regeneration of potato plants is described in great detail by Shepard (6a). Many color illustrations portray the steps of the fusion process in an enlightening manner.

Radiation-Induced Mutations

Jinks' work (7) with *Nicotiana rustica*, a common variety of tobacco plant, concerned the effects of γ-irradiation upon plant pollen. It was found possible to transfer specific paternal characteristics into a maternal background by cross pollenation with irradiated pollen. The transfer of such characteristics by this method is much faster than could be achieved in conventional breeding. In a general review of gene transfer in plants, Davies (8) compares the results of Jinks and data from some other reports. A possible speculative explanation would be than ionizing radiation breaks the chromosomes into fragments, probably randomly. Then, transfer of the fragments, which now represent altered gene sequences, to the egg results in incorporation of the fragmented gene sequences.

Agricultural Microbiology

A thought-provoking article by Brill (9) elaborates on the interplay of the microbial fauna in the soil with the flora growing therein. His opening statement is worth quoting: "Introducing new genes into crop plants by recombinant DNA methods is difficult and not in immediate prospect. Much progress can be made, however, by manipulating the microorganisms that live with plants." An article by Rhodes-Robert (9a) covers the interactions between bacteria and plants in the root environment. All aspects of symbiotic relationships are treated in depth.

A procedure which could effectively confer beneficial properties to the plant envisions the culturing of tailored microbes–created in a fermentation type process–and their introduction to the plant soil, or interaction with isolated plant cells in a nutrient solution. Considerable emphasis is of course, placed on the possibility that nitrogen fixation can thus be incorporated into a plant-microbe symbiosis.

Bacterial candidates are those of the *Rhizobium* genus, which normally infect root hairs, the bacterium *Axotobacter vinelandi* (under investigation by Brill's group), Ti plasmids in *Agrobacterium tumefaciens*, and *Klebsiella pneumoniae nif* genes.

An example of the introduction of nitrogen-fixation (nif) genes into the genome of a yeast is illustrated in Figure 1, which is a schematic of an experiment conducted by Szalay's group at Cornell University. However, the nitrogen-fixing proteins were not expressed in the yeast. The example is quoted as it entails the recombinant procedures which will have to be used in such innovative, albeit speculative attempts, where bacterial and plant plasmids, as well as chromosomal DNA are subjected to cleaving, rejoining, and introduction into a host cell in the hope that a desirable function will ultimately be expressed by a recombinant chromosome. It should be noted that this procedure could be included under recombinant methods.

Agrobacterium rhizogenes, a plant infectant closely related to *Agrobacterium tumefaciens*, was studied by a group of investigators (10) in France, who were interested in the phenomenon of "hairy root disease." Just as in crown gall disease, the *A. rhizogenes* bacterium induced infected plant tissue to produce the novel metabolites, the "opines." The virulent infections are conferred by large plasmids similar to the Ti plasmids, where a specific sequence in the genome-transferred DNA (C-DNA) is incorporated into the host plant's nuclear DNA. The investigation showed that the *rhizogenes* bacterium acts in the same manner as the Ti plasmid and the infectious plasmid was labeled as "Ri-plasmid," that is, "root inducing virulence plasmids." Cleaving and sequence studies showed that the T-DNA is located in the pRi 8196 plasmid. The behavior of the Ri organisms suggests that it is a suitable vector in recombinant engineering.

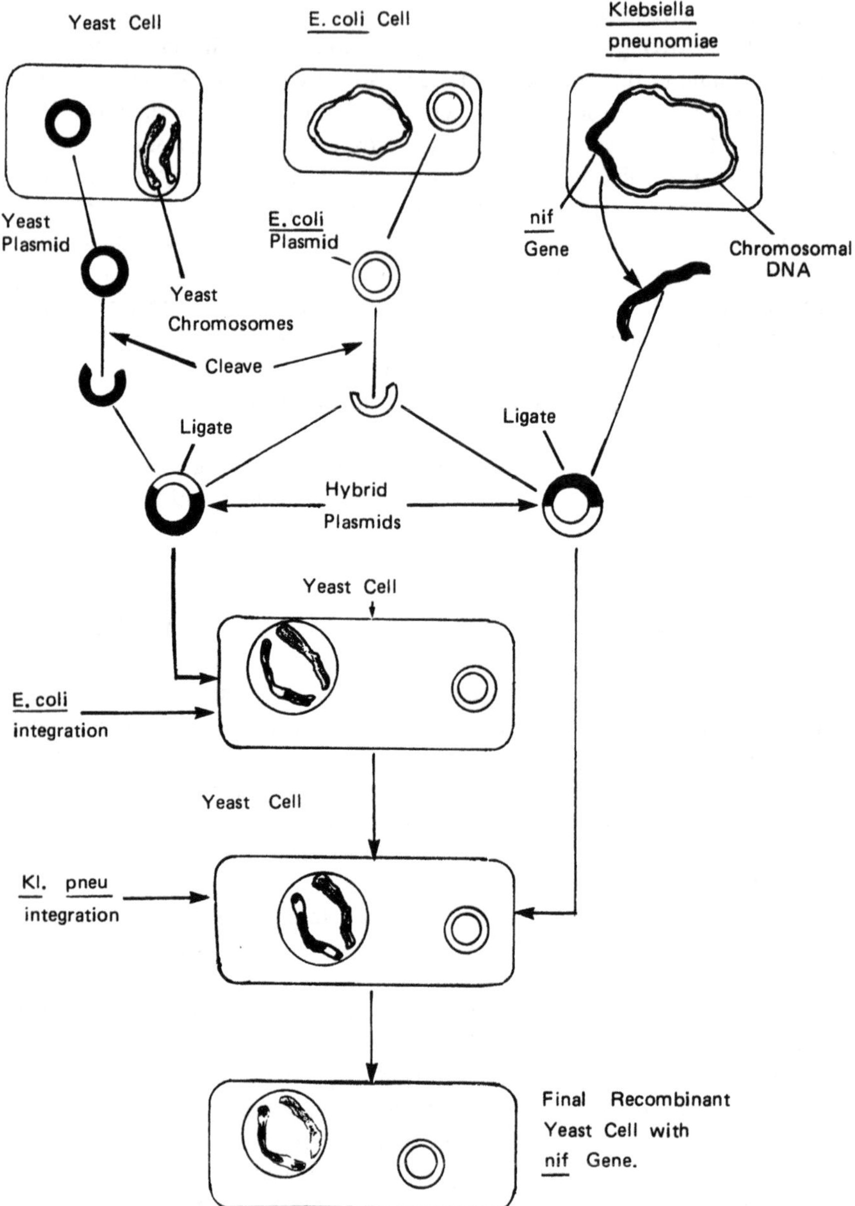

FIGURE 1 Insertion scheme for nitrogen fixation gene. The two-stage process diagrams the insertion of the nif gene into the genome of a yeast (Ref. 9).

Recombinant Procedures

The subject of genetic manipulation in plants was reviewed comprehensively by Cocking et al. (2). Cocking points out that there are two major experimental approaches to attempt the insertion of foreign DNA into plant cell genomes. One method would involve the combination of plant DNA sequences with foreign (exogenous) DNA, so that a homologous sequence is created between that plant DNA (a host DNA) and the vector DNA. The other method would use tumor-inducing (Ti) plasmids from *Agrobacterium tumefaciens* as a vehicle to introduce foreign DNA into cells of higher plants. The possible procedures to accomplish the DNA alterations are treated in detail. Also, covered are the isolation of plant genes and somatic cell fusion (see previous section on protoplast fusion).

Plant modification by recombinant DNA technology follows the normal procedures. Thus, plasmids from the plant or more likely from bacterial cells can be isolated and altered genetically. Reintroduction into the cell can, but not always will, lead to the expression of the genetic modification.

Zein Modification

The work of Burr (11) on Zein, a protein in the corn, can be considered as an example of recombinant procedures, specifically the use of enzymatic strand cutting and RNA to cDNA conversion. Burr and co-workers prepared double-stranded complementary DNA (cDNA) from purified Zein mRNA, and inserted the cDNA into the bacterial plasmid pMB9 by using the A-T tailing method. Twenty clones were examined and 18 clones were found to contain translatable Zein mRNA.

Plant Chromosomal DNA

Another example of the use of genetic engineering techniques in plant research is the work of Bedbrook and Gerlach (12) dealing with the elucidation of chromosomal DNA in plants. Such DNA contains a large proportion of repeated sequences (non-sense DNA, introns). The repeat structures amount to at least 70% of the total chromosomal DNA.

Bedbrook and Gerlach used molecular cloning to isolate repeated DNA sequences. So, the eukaryotic plant DNA when treated with restriction endonucleases (RT) gave size-fractionated digests. The repeat sequence appeared as bands on electrophoresis gel. Because RT will not cut repeat sequences when the repeat unit does not contain restriction sites for the RT, the uncut sections amount to an enrichment of repeats. For instance, the DNA fractions obtained from *Eco*RI-digested wheat DNA was excised from the gel and cloned in plasmid pAC184 in a *E. coli* host strain.

Cloning experiments gave high transformation efficiency and relatively large quantities for DNA sequencing. In an informative appendix the detailed procedure is given for isolating high molecular weight DNA from cereal embryos.

Ti Plasmids

The use of the tumor-inducing plasmids occurring in the *Agrobacterium tumefaciens* was already illustrated in the discussion of Agricultural Microbiology. Indeed, there is tremendous interest in using the Ti plasmids to introduce new genes into plant DNA. As a matter of fact, the potential utilization of this plasmid is being investigated on a worldwide scale.

A detailed study is reported by Hooykaas, et al. (13, The Netherlands). After delineating the physiology of tumor induction, the authors present information of available strains and their suitability to undergo genetic transformation. Attempts to eliminate the gene for crown gall induction, an undesirable feature of Ti plasmids, so far have had only limited success (14). Another limitation of the usefulness of *Agrobacter* is that the hosts which it can infect are limited to dicotyledons, that is, plants whose seeds germinate to form two leaves. This eliminates Ti plasmid use in cereal grains and many other economically important plants.

An interesting phenomenon associated with Ti plasmids is the formation of unusual amino acid derivatives in the tumor structures. These compounds are called opines and most of them are amino acid condensates to pyruvate.

A method for insertion of genetic material into a specific site of the T-DNA sequence in a Ti plasmid was reported by Matzke and Chilton (15). In a three-step recombinant procedure they inserted two pBR322 fragments into a pRK290 plasmid, also containing kanamycin resistance markers, and transformed into an *Agrobacterium tumefaciens* strain having a wild-type Ti plasmid. Some of the engineered strains showed greatly decreased tumor production in tests on three plant species. Similarly, Krens et al. (16) accomplished in vitro transformation of plant protoplasts with Ti-plasmid DNA.

Sunbean

Reports (17) publicizing the creation of "sunbean," a hybrid of French bean and sunflower, have appeared only as news articles. It is reported that the hybridization was accomplished by means of a crown gall bacterium. This success is an example of the potential existing to create new and hopefully improved plant varieties.

Tobacco Mosaic Virus (TMV)

A significant finding was reported by Devash et al. (18) on the inhibition of TMV by the oligonucleotide (2′-5) oligoadenylate. This compound is induced in animal cells that have been treated with monkey and human interferon. The proliferation of TMV was reduced by as much as 90%. The authors speculate that plants and animals may have a common pathway for virus resistance.

The most recent findings on potential recombinant applications to agricultural problems were reported at a symposium in May 1982 (19). Advances in the successful use of Ti plasmids, studies on soybean genes, viroids of potato spindle disease, composition of cereal plant DNAs, and so on were presented. A quotation by Richard Flavell (Cambridge, England) highlights the prevailing consensus on the genetic engineering in improving crop plants:

> Genetic engineering of improved crop plants will be much more difficult (than presently available methods). You are only putting a single gene into bacteria. But many genes are involved in determining plant characters. . . . The time scale for genetically engineering plant improvements is not a few years but longer.

An informative and challenging composition on the plant sciences is the subject of the May/June 1982 issue of *Mosaic*. In particular an article by W. A. Check gives an overview of genetic engineering applied to the plant gene (20).

In an International Conference on Genetic Engineering, sponsored by Battelle Memorial Institute (at Reston, VA in 1981) a great variety of subjects were covered. One of the major aspects dealt with agricultural applications. Topics in nitrogen fixation, genetics of photosynthesis, new hybrid plants, enhancing protein quality and quantity, *Agrobacterium tumefaciens* tumor-inducing plasmids, herbicide resistance, and current research trends in agricultural genetic engineering were published in a volume of the proceedings (21). Two recent volumes of interest are one that deals with the origins of chloroplasts (22) and another text that covers a variety of new techniques in plant cell culture (23).

The continued high interest in the use of genetic engineering techniques, both recombinant and nonrecombinant, to plant beneficiation is exemplified by three articles in a May 1983 issue of the *New Scientist*. There, Tudge (24) writes on worldwide successes in crop culturing to gain plant resistance to frost and brackish water; Branton and Blake (25) report progress in coconut palm reproduction via tissue culturing; and Smith (26) deals with gene bank establishments for a multitude of potato species.

REFERENCES

1. Keller, EF. McClintock's maize. *Science 81 2* (8). 55–58 (1981).
2. Cocking, EC et al. Aspects of plant genetic manipulation. *Nature* 293(5830): 265–270 (1981).
3. Wang, W-Y. Plant genetics, University of Iowa, (Personal communication).
4. Fox, JL. Plant molecular biology beginning to flourish. *chem Eng. News 59* (25):33–44 (1981).
5. Blair, JG. Test tube gardens. *Science 82* 3 (1):70–76 (Jan/Feb. 1982).

6. Farkas, GL, Ferenczy, L, (Eds). (1980). Advances in protoplast research. *Proc. 5th Entl. Protoplast Symp.*, July 9–14, 1979. Pergamon Press, New York.
6a. Shepard, JF. The regeneration of potato plants from leaf-cell protoplasts. *Sci. Am.* 246 (5):154–166 (1982).
7. Jinks, JL et al. Gene transfer in *Nicotiana rustica* using irradiated pollen. *Nature* 291 (5816):586–588 (1981).
8. Davies, R. Gene transfer in plants. *Nature* 291(5816):531–532 (1981).
9. Brill, WJ. Agricultural microbiology. *Sci. Am.* 245 (3):199–215 (1981).
9a. Rhodes-Roberts ME, Skinner FA. (Eds). (1982). *Bacteria and Plants* Academic Press, New York.
10. Chilton M-D et al. Agrobacterium rhizogenes inserts T–DNA into genomes of host plant root cells. *Nature* 295(5848):432–434 (1982).
11. Burr, B. (1980). The use of recombinant DNA methodology in approaches to crop improvements: The case of Zein. *Genetic Engineering, Principles and Methods*, Setlow, JK, Hollaender, A Eds., Plenum Press, New York, Vol. 2, pp. 21–29.
12. Bedbrook, JR, Gerlach, WL. (1980). Cloning of repeated sequence DNA from cereal plants. In *Genetic Engineering*, Setlow, JK, Hollaender, A. (Eds.), Plenum Press, New York, Vol. 2, pp. 1–19.
13. Hooykaas, PJJ et al. (1981). *Agrobacterium* tumor inducing inducing plasmids: potential vectors for the genetic engineering of plants. In *Genetic Engineering*, Academic Press,London, Vol. 1, pp. 151–179.
14. Neidle, S. (1981). *Topics in Nucleic Acid Structure*. John Wiley & Sons, New York.
15. Matzke, AJM, Chilton, M-D. Site specific insertion of genes into T–DNA of the *agrobacterium* tumor-inducing plasmid: an approach to genetic engineering of higher plant cells. *Mol. Appl. Genet.* 1 (1):39–49 (1981).
16. Krens, FA et al. *In vitro* transformation of plant protoplasts with Ti-plasmid DNA. *Nature* 296(5852): 72–74 (1982).
17. Anonymous. Project sunbean: plant gene transfer. *Sci. News* 120 (2):23 (1981).
18. Devash, Y, Briggs, S, Sela, I. *Science* 216 (4553): 1415–1416 (1982).
19. Marx, JL. Ti Plasmids as gene carriers. Report on Symposium" Genetic Engineering: Applications to Agriculture. 16 6o 19 May, 1982, Beltsville, MD.
20. Check, WA. Engineering the botanical Gene. *Mosaic* 13 (3):19–24 (1982).
21. *Proceedings 1981 Battelle Conference on Genetic Engineering.* (Ed.) Keenberg, M. Battelle Seminars and Studies Program Battelle Institute, Columbus, Ohio.
22. *On the Origins of Chloroplasts* Schiff, JA. (Ed.). (1982). Elsevier/North-Holland, New York

23. Fiechter, A. (Ed.). *Advances in Biochemical Engineering* Plant Cell Cultures II (1980). Springer Verlag, New York.
24. Tudge, C. The future of crops. *New Sci.* 98(1359):547–553 (1983).
25. Branton, R, Blake, J. A lovely clone of coconuts. *New Sci.* 98(1359):554–557 (1983).
26. Smith, N. New genes from wild potatoes. *New Sci.* 98(1359):558–565 (1983).

10 Genetic Engineering Activities

A cursory perusal of scientific papers dealing with genetic engineering research and production might leave the impression that only a limited number of prestigious institutions are active. However, if news items as well are screened for information, it becomes evident that a regular "Who's Who" coterie of the pharmaceutical and chemical industries is engaged in this area. An example is a listing of 30 participants on page 70 of the January 1982 issue of *Bio Science*.

One indicator of trends in a manufacturing industry is the existence of *trade journals*. So, it is significant that the first such publication premiered in January 1981 as the *Genetic Engineering News* (1). A very comprehensive report on genetic engineering and biotechnology firms through 1981 is available from Noyes Data Corp. Some 250 firms are listed with essential details. The substantial price is quoted as $100 (2). A recent survey in the February 24, 1982, issue of *Chemical Week* is of more limited scope but quite informative. The potential success of the many commercial small enterprises that have sprung up is questioned and a considerable number of washouts is foreseen. A relatively new monthly publication is briefly described in Chemtech (2a). It is entitled *Telegen* and intended to cover worldwide *biotechnology and genetic engineering* as an intelligence-gathering project. One wonders, however, if the $1,200 yearly price tag will enable many investigators to subscribe, and so it is probably directed to a "company" clientele.

FIELDS OF ENDEAVOR

Present-day emphasis in the use of genetic engineering is concentrated in the more glamorous aspects of the medical area such as production of insulin, interferons, human growth hormone, somatostatin, etc. But the potential benefits to other areas may surpass in importance the production of pharmaceuticals. For instance, while medicine is essential in sickness, food is needed at all times by all people. Consequently, great advances will arise in agricultural genetic engineering.

Considerable speculation has been exercised in the possible utilization of recombinant methods for the manufacture of chemicals. Developments in this area will be strictly dependent on whether such methods can be more profitable than conventional chemical manufacturing. Some of the envisioned creations in the agricultural area have been covered in the chapter on "plant genetics." Although there have been speculative assertions that genetic engineering methods may find their widest application in the large-scale production of chemicals, there is no documented evidence of any substantial progress. A review of an Office of Technology Assessment report and some information gleaned from industry sources (3) present an interesting listing of a multitude of organic chemicals which might be long-range candidates for applied genetics manufacturing (Table 1). The opinions of personnel at Genex and MIT agreed on only 8 compounds, however.

The information which becomes available almost daily precludes a comprehensive coverage of the many new developments, however, interesting and even startling they might be. An attempt has been made to select some prominent examples, which have either been carried to an advanced stage, or would be considered to be particularly promising.

SOMATOSTATIN

Somatostatin (SOM) is a human peptide (a proteinaceous molecule). As such it is a hormone produced in the hypothalamus at the base of the brain, from where it is transported via the blood stream to the pituitary gland. There it acts as a control to inhibit the release of insulin and human growth hormone. It is a rather small polypeptide, consisting of only 14 amino acids requiring 42 base pairs for the code. Somatostatin is the first human polypeptide which was produced in bacterial cells.

The diagramatic procedure for SOM synthesis is presented in Figure 1 (4). From the known amino acid sequence a corresponding nucleotide sequence was synthesized. An additional 10 base pairs were inserted at the ends to create "sticky ends" which would facilitate insertion into a modified pBR 322 plasmid. The *lac* operator control region and a large part of the enzyme beta-galactosidase gene in the plasmid served to assure the expression of the galactosidase gene. As shown, SOM gene was inserted next to the enzyme gene. After introduction of the plasmid into *E. coli* the hormone was produced with a short peptide tail at

TABLE 1 Production by Applied Genetics: Chemicals That Might be Made Biologically Within 20 Years

	Genex	MIT
Acetaldehyde		X
Acetic acid	X	X
Acetoin		X
Acetone		X
Acetylene		X
Acrylic acid	X	X
Adipic acid	X	X
Aniline	X	
Aspirin	X	
Benzoic acid	X	
bis(2-ethylhexyl)adipate	X	
Butadiene		X
Butanol		X
Butyl acetate		X
Butyraldehyde		X
Cinnamaldehyde	X	
Citronellal	X	
Citronellal	X	
Cresols	X	
Dihydroxyacetone		X
Diisodecyl phthalate	X	
Dioctyl phthalate	X	
Ethanol	X	X
Ethanolamine	X	
Ethyl acetate		X
Ethyl acrylate		X
Ethylene		X
Ethylene glycol	X	X
Ethylene oxide	X	X
Formaldehyde		X
Geraniol	X	

TABLE 1 (Continued)

	Genex	MIT
Glycerol	X	X
Isobutylene	X	
Isoprene		X
Isopropanol		X
Itaconic acid	X	
Linalool	X	
Linalyl acetate	X	
Methane	X	
Methanol		X
Methyl acrylate		X
Methyl ethyl ketone		X
Nerol	X	
para-Acetaminophenol	X	
Pentaerythritol	X	
Phenol	X	
Phthalic anhydride	X	
Propionic acid	X	
Propylene		X
Propylene glycol	X	X
Propylene oxide		X
Sorbic acid	X	
Sorbitol	X	
Styrene		X
alpha-Terpineol	X	
alpha-Terpinyl acetate	X	
Vinyl acetate		X

Source: Studies by Genex Corp. and the Massachusetts Institute of Technology for the Office of Technology Assessment, U.S. Congress.
Courtesy *Chem. Week* (Ref. 3).

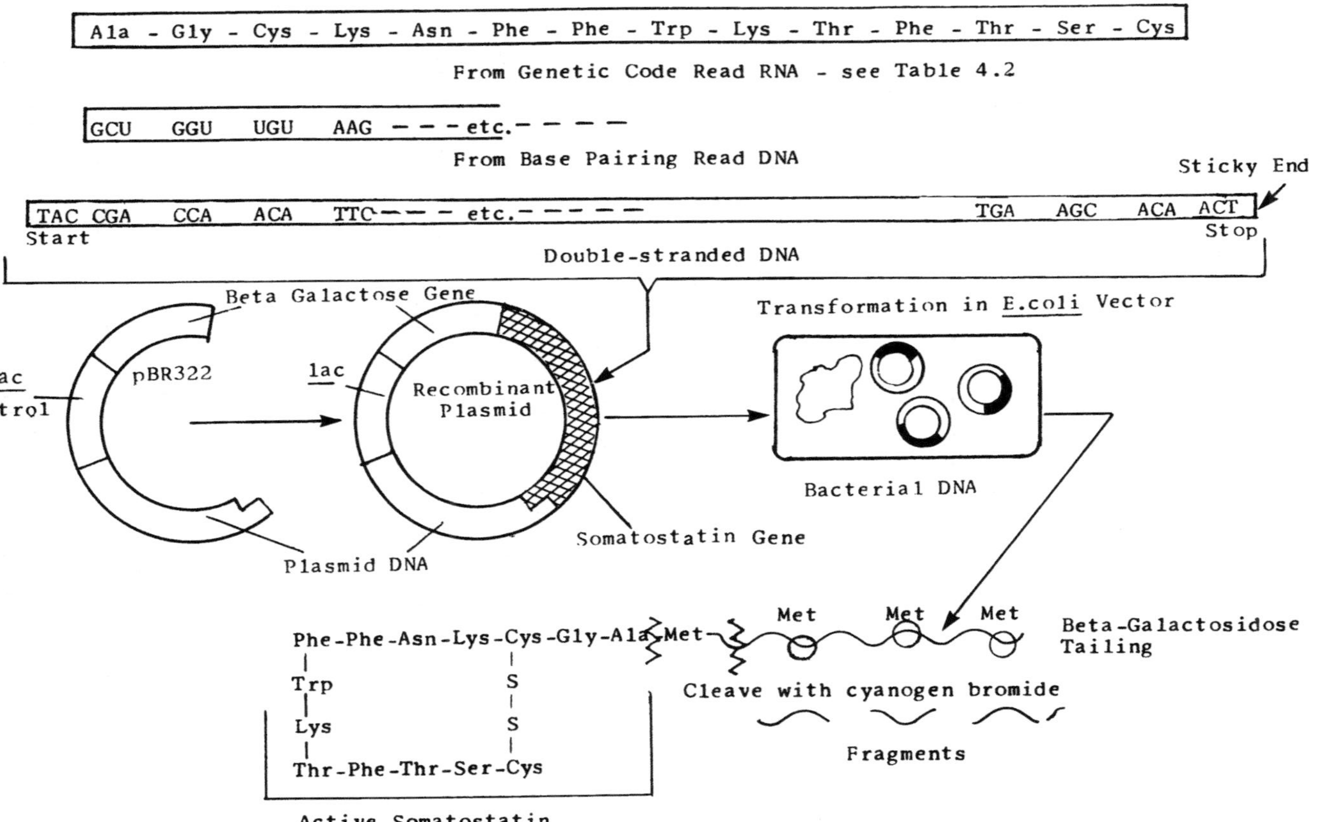

FIGURE 1 Synthesis of somatostatin. From the 14 amino acids sequence the mRNA composition is read using the genetic code. The DNA Gene is obtained from base pairing and inserted into a pBR322 plasmid. Cloning in *E. coli* factor yields a somatostatin with BG and Met tailing. Cleaving at Met-Ala position gives active somatostatin (Ref. 4).

the end consisting of alternate methionine and enzyme fragments. The hormone itself was then produced in its pure form by removing the peptide tail with cyanogen bromide treatment. The synthesized SOM product has been shown to be identical to the natural hormone. While productivity on cloning was only marginally good, stated to be 10,000 SOM molecules per cell, this production rate was considered to be encouraging.

A very detailed procedure, including the formation of intermediate stage plasmids from the pBR322 type, is described by Old and Primrose (5). The steps cover the insertion of the *lac* control region to give a pBH10 plasmid, the determination of the orientation of the *lac* insert, the removal of one *Eco*RI site (enzyme cleavage site) to form a pBH20 plasmid, and insertion of the SOM gene and the β-glactosidase gene in a 3-step procedure. Each step is well illustrated by lucid diagrams. The authors also stress the important fact that the procedure included the creation of a functional gene from chemically synthesized gene fragments.

INSULIN

The subject of insulin is much in the foreground of academic and public thinking. As a result, a plethora of publications is to be expected. A report by Losse and Mätzler (6) is noteworthy for its coverage of a multitude of modified insulins arrived at by functional substitutions. The modified insulin molecules are of interest because of potentially greater specificity in action as well as possible extension of utility.

There are many pharmaceutical products under investigation. In all respects, however, the production of *human insulin* by recombinant techniques is the most advanced undertaking. News items (7) report that the first plant using genetic engineering techniques was started up in England. The plant is operated by Dista Products, a subsidiary of Eli Lilly and Co. Biosynthetic human insulin (BHI) is manufactured in multipound amounts. The product is already for sale in England and has been approved by FDA in the USA under the tradename Humulin.

Synthesis of Human Insulin

Insulin for human medication had been obtained for many years from pork and beef pancreas (porcine and bovine insulins). They are still the products presently on the market (see Fig. 2a (8), which shows the structure of beef insulin). The so-called *A* chain contains 21 amino acids and an internal disulfide bridge. The longer *B* chain is made up of 30 amino acids. The two chains are crosslinked by two disulfide bridges. The bovine insulin and human insulin differ in only two amino acids. The alanines in bovine are replaced by threonine in both chains, and one valine (No. 10) in the *A* chain becomes isoleucine in the human molecule. Figure 2b pictures the insulin precursor "proinsulin" of the pork variety. Comparisons of insulins from animal sources and human insulin are listed in Figure 3.

Chemical Synthesis of Insulin Genes

Crea et al. (9) first reported on the chemical synthesis of insulin genes in 1978. The synthesis was accomplished by preparing 29 oligodeoxyribonucleotides (dimers, trimers, and tetramers) using an improved triester method (Fig. 4). These fragments were ligated in the proper sequence to obtain the DNAs for the *A* and *B* human insulin chains (Fig. 5).

COMMERCIAL SYNTHESIS OF HUMAN INSULIN

The synthesis of human insulin is being undertaken by Eli Lilly and Company (Indianapolis, IN) utilizing developments in genetic engineering by Genentech, Inc. (San Francisco, CA).

The Lilly process consists of the production of the two insulin chains and subsequent combination of the chains through disulfide bonds. This is called the two-pot process. The *A* chain contains 21 amino acids and an intra chain –S–S bond. The *B* chain contains 30 amino acids, but no internal –S–S bond (see Fig. 3).

The vector for gene insertion is a plasmid from *Escherichia coli*. The host cell for the modified plasmids also is *E. coli*, the K12 strain. While the complete details of Lilly's procedure were not published as of this writing (Sept. 1982), a company bulletin (8) and a published article (10) indicate that the following steps are used:

1. From the genetic code in Table 2, the mRNA sequence is read. Then the single-strand DNA sequence is established using the base-pairing rule. Finally, fragments are constructed by synthesis of oligonucleotides.
2. The fragments are ligated in proper sequence and hybridized to yield double-stranded DNA (ds-DNA).
3. ds-DNA sequences are inserted into a plasmid (see Figs. 6 and 7).
4. Figure 8 depicts isolation of "active" plasmids, using TET^R and AMP^R genes of the plasmids to obtain viable clones. Subsequent density gradient separation isolated DNA fragments and plasmid DNA. Cleaving of a desired sequence from foreign DNA, insertion into plasmid vector and recombination are illustrated in Figure 9.
5. Combining *A* and *B* chains to final product: Lilly's procedure for combining the *A* and *B* chains by means of disulfide crosslinking and

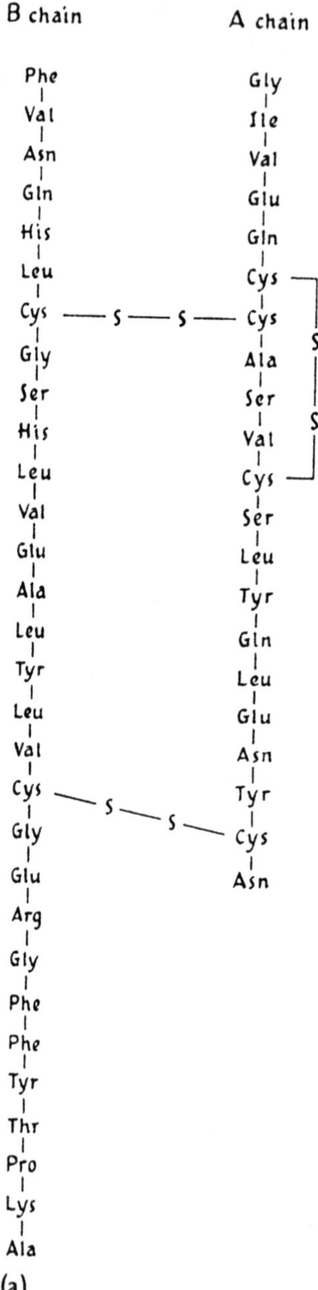

FIGURE 2 Structures of insulins. (a) Beef Insulin. Insulins consist of two amino acid chains: Short A chain: 21 AAs; Long B chain: 30 AAs.

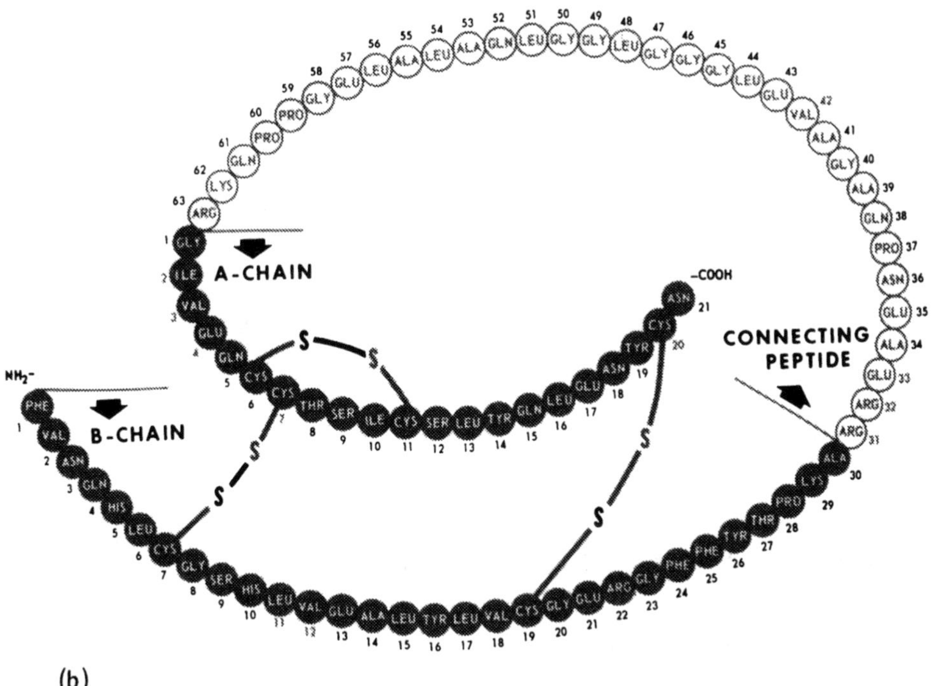

(b) Porcine Insulin. Structure of proinsulin, which becomes insulin after splitting off the polypeptide chain containing 33 AAs. Courtesy Eli Lilly and Co. (Ref. 8).

introduction of a disulfide bridge on the *A* chain was reported by Chance (10). It is obvious that the attainment of proper locations of these bonds is a complex problem. The potential number of combinations which could occur in all three of the disulfide locations is indicated in Figure 10 and the desired bonding scheme is as shown in Figure 11.

The two insulin chains are obtained from separate *E. coli* fermentations (culturing) with the *A*- or *B*-chain genes linked to a "leader DNA" (e.g., tryptophan synthetase gene), through a methionine codon (see Fig. 12). The methionine ends are removed by reaction with cyanogen bromide. The disulfide creation is via S-sulfonates which are purified by chromatography. Chemical combination of the sulfonated chains follows.

Details of Combination Procedure

Indicated in Figure 11 is the amount of DTT (dithiothreitol) added to give the desired SH: SSO_3^- molar ratio of 1.2. The DTT is dissolved in a 0.1 *M* glycine

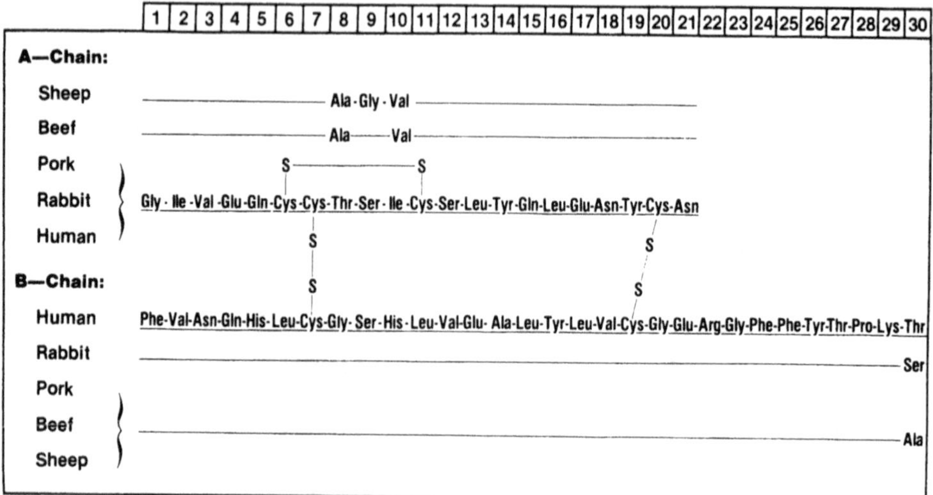

FIGURE 3 Insulin structures. Structures of animal insulins are compared with human insulin. Courtesy Eli Lilly and Co. (Ref. 8).

buffer at PH 10.5, and quickly added to a chilled chain solution (5–20 mg/ml) which contains *A* and *B* in a ratio of 2:1 by weight of s-sulfonated chains. The mixture is stirred for 24 h at 4°C in an open vessel to permit air oxidation.

Yield is ~ 60% of theory based on amount of *B* chain used. Unused products are recycled. DTT = dithiothreitol readily oxidizes to a stable six-membered ring and does not participate further in the combination reaction, and oxidized DTT is easily removed from the combination solution in the subsequent insulin purification steps.

A pictorialized version of the overall process DNA→mRNA→protein synthesis→proinsulin→insulin is shown in Figure 13. Intermediate reactions, such as tRNA function, are described in Chapter 4.

HUMAN GROWTH HORMONE (HGH)

Before recombinant techniques could be considered for the production of HGH, the only sources of supply were the pituitary glands from human cadavers. Even then the purity of extracted hormone was beset by the presence of modified forms of HGH, as well as fragmented structures, so that a rather high degree of inhomogeneity was encountered. The promising use of HGH, also called "somatotropin," to alleviate genetic dwarfism, and its apparent benefits in healing of fractured bones, promoting skin regeneration in burn cases, and possibly in

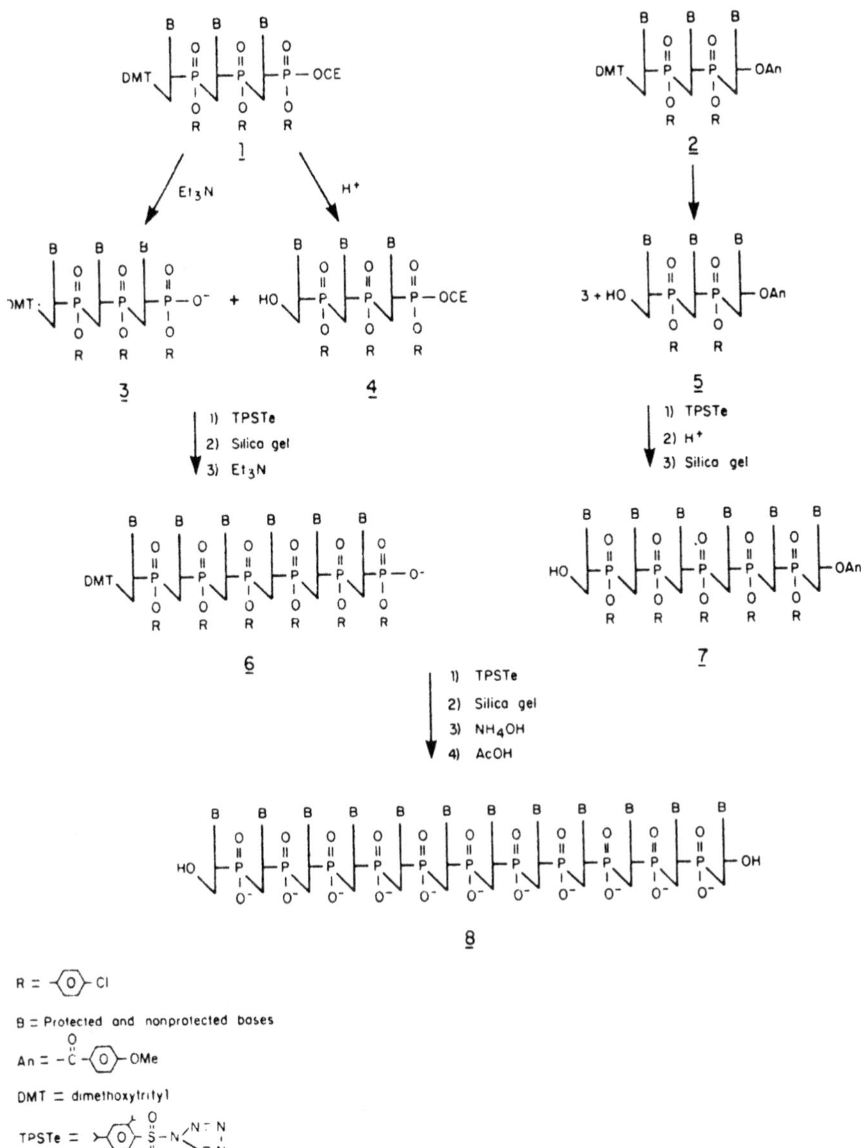

FIGURE 4 Synthesis procedure for human insulin. Reaction mechanism used to prepare short segments for subsequent ligation. Courtesy Natl Acad. Sci., USA (Ref. 9).

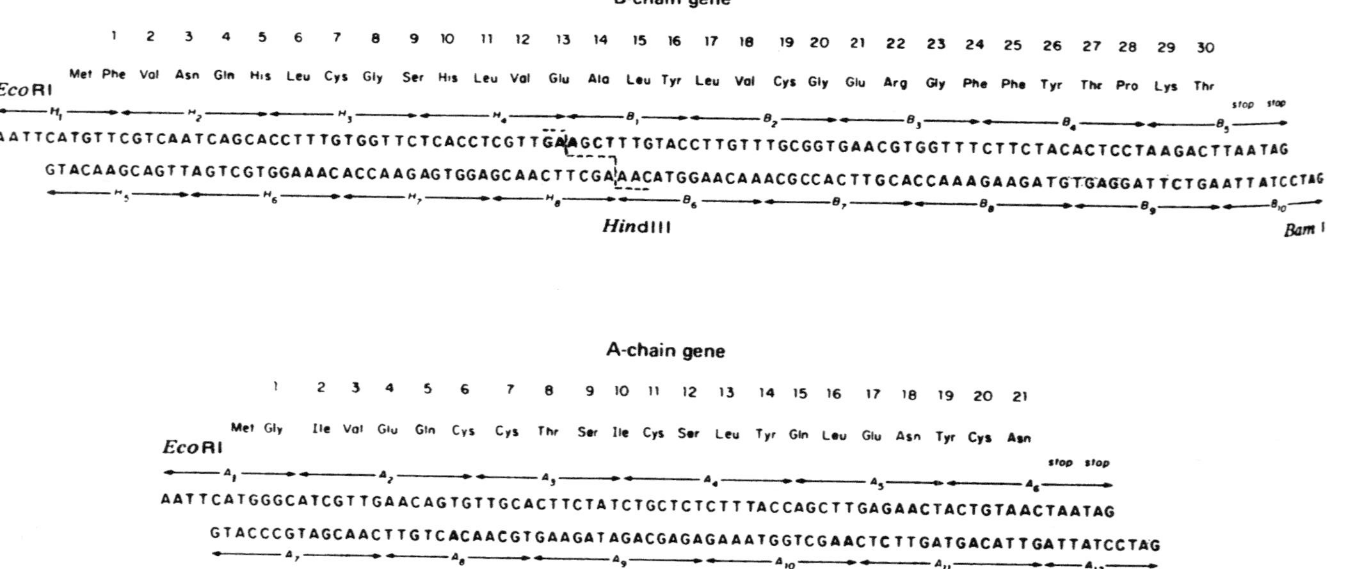

FIGURE 5 Human insulin structure. Amino Acid to RNA to DNA conversion. Methionine end is removed to obtain final structure. Disulfide bridges will complete insulin molecule. Courtesy Nat. Acad. Sci., USA (Ref. 9).

TABLE 2 Commercial Synthesis of Human Insulin

	U		C		A		G	
U	UUU	Phe	UCU	Ser	UAU	Tyr	UGU	Cys
	UUC	Phe	UCC	Ser	UAC	Tyr	UGC	Cys
	UUA	Leu	UCA	Ser	UAA	Ochre	UGA	Umber
	UUG	Leu	UCG	Ser	UAG	Amber	UGG	Trp
C	CYY	Leu	CCU	Pro	CAU	His	CGU	Arg
	CUC	Leu	CCC	Pro	CAC	His	CGC	Arg
	CUA	Leu	CCA	Pro	CAA	Gln	CGA	Arg
	CUG	Leu	CCG	Pro	CAG	Gln	CGG	Arg
A	AUU	Ile	ACU	Thr	AAU	Asn	AGU	Ser
	AUC	Ile	ACC	Thr	AAC	Asn	AGC	Ser
	AUA	Ile	ACA	Thr	AAA	Lys	AGA	Arg
	AUG	Met	ACG	Thr	AAG	Lys	AGG	Arg
G	GUU	Val	GCU	Ala	GAU	Asp	GGU	Gly
	GUC	Val	GCC	Ala	GAC	Asp	GGC	Gly
	GUA	Val	GCA	Ala	GAA	Asp	GGA	Gly
	GUG	Val	GCG	Ala	GAG	Glu	GGG	Gly

Note: U = uridine, an RNA nucleotide that is replaced with thymidine in DNA. Ochre, Amber, and Umber are stop codons that signal the end of a gene.
Courtesy Eli Lilly and Co. (Ref. 8).

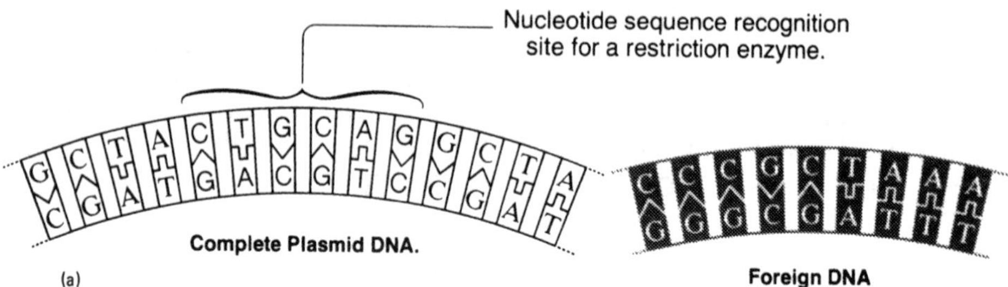

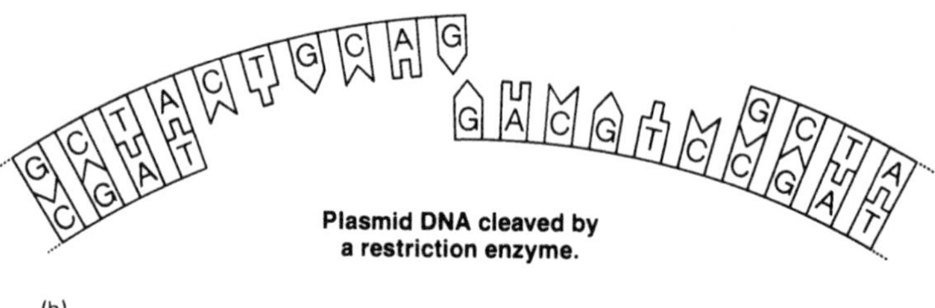

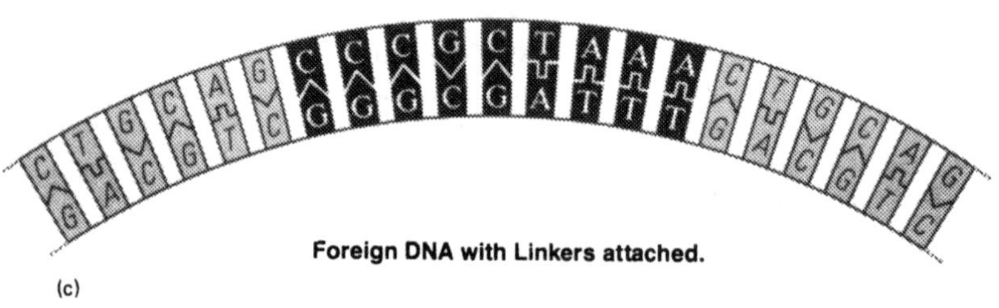

FIGURE 6 Recombinant procedure. Plasmid *E. coli* cleavage and linker attachment to foreign DNA.

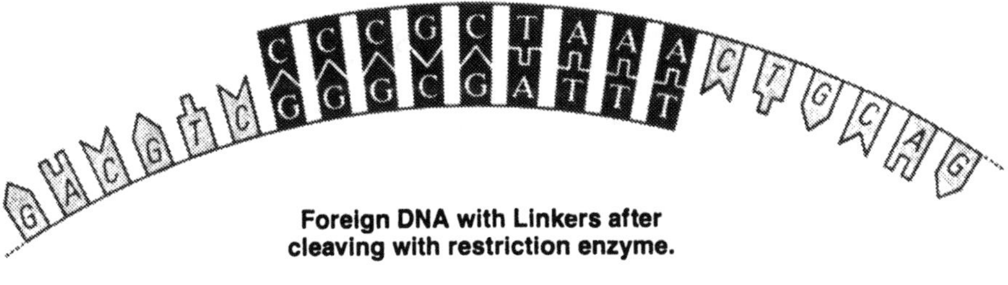

(a)

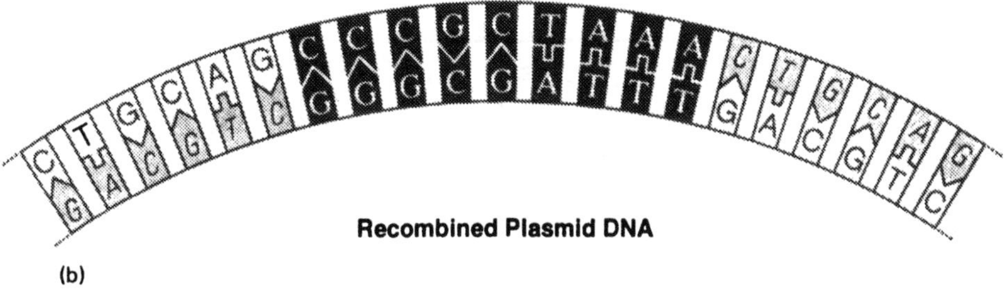

(b)

FIGURE 7 Recombinant procedure. Linker-end preparation and insertion into plasmid. Courtesy Eli Lilly and Co. (Ref. 8).

therapeutic treatment of ulcers, stimulated widespread interest in developing recombinant procedures for its synthesis.

HGH is a eukaryotic protein containing 191 amino acids. Aharonowitz and Cohen (10a) presented a recombinant scheme. Figure 14 is a simulation of the procedure. While not specifically stated, one might infer that the recombinant steps were developed by cooperation of Genentech and Kabi Gen AB (Sweden) and is patterned on the work reported by Goeddel et al. (11). As indicated, human pituitary glands were processed to extract the mRNA of HGH and make a cDNA copy by reverse transcriptase. Treatment with a restriction endonuclease eliminated an oversize front, while retaining the code for the amino acids number 25 to 191. The condensed sequence of the gene for numbers 1 to 24 was prepared from 12 fragments of synthetic nucleotides through an intermediate coalescence to 3 fragments into the condensed gene. Further ligation yielded the complete gene with one sticky end having the initiation codon ATG and a blunt end. A pBR322 plasmid, with the *lac* operator modified

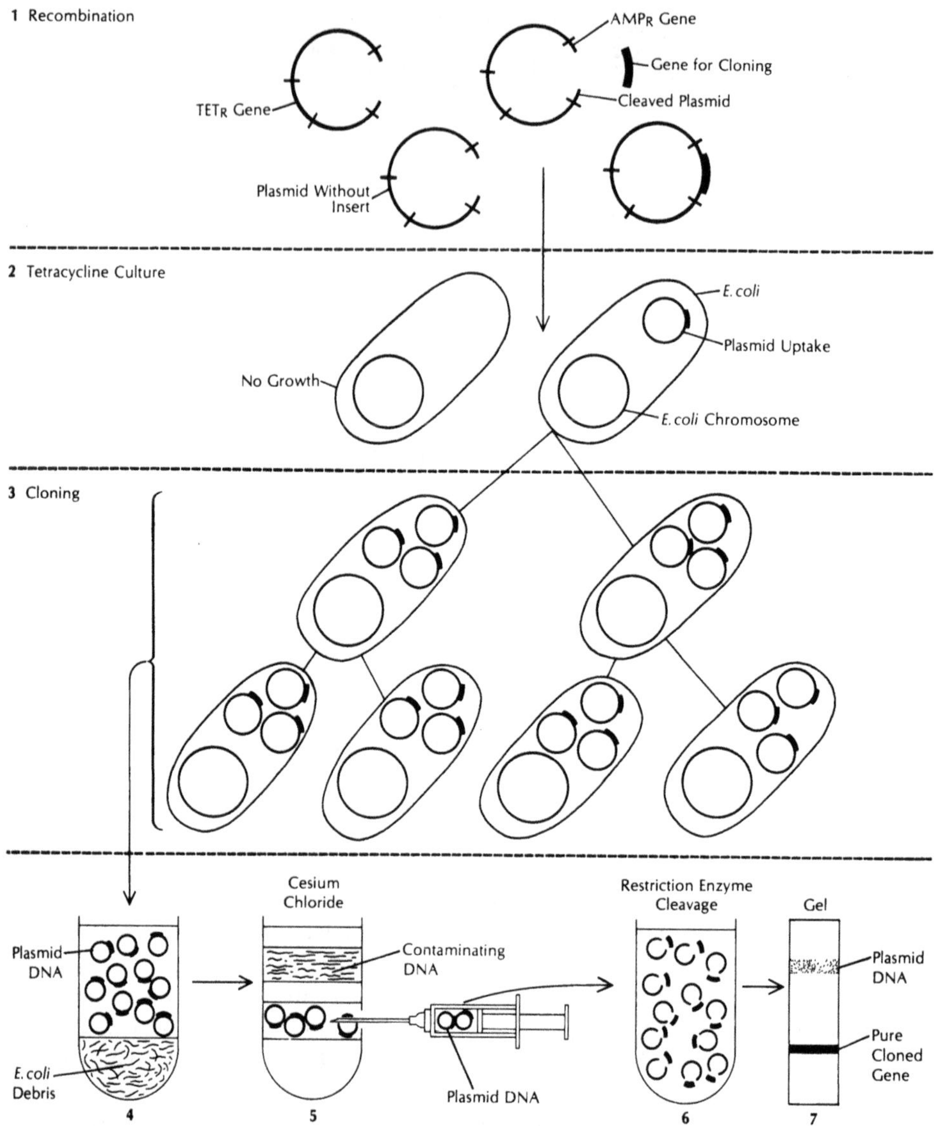

FIGURE 8 Recombinant scheme. Plasmid preparation, gene insertion, Tet and Amp screening and cloning. Final plasmid isolation by density gardient separation. Courtesy Eli Lilly and Co. (Ref. 8).

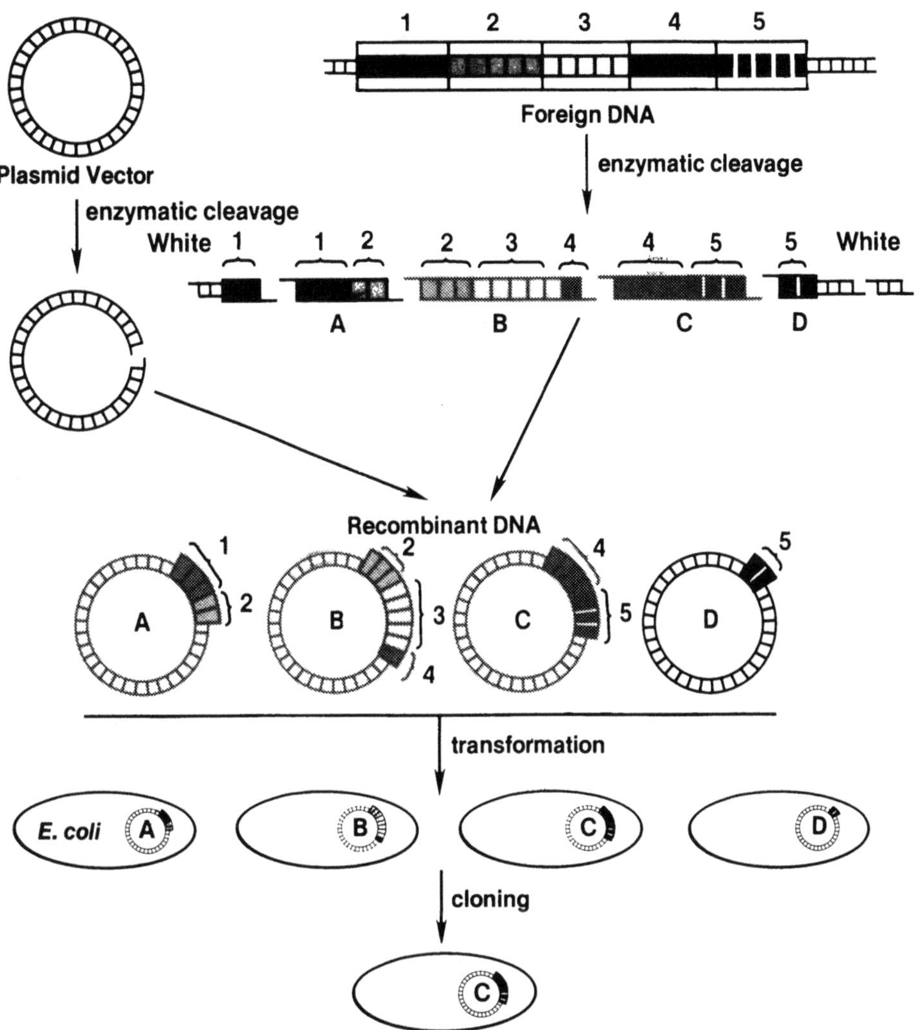

FIGURE 9 Insertion and cloning. Sequential diagram elaborating steps in Figures 6 and 7. Courtesy Eli Lilly and Co. (Ref. 8).

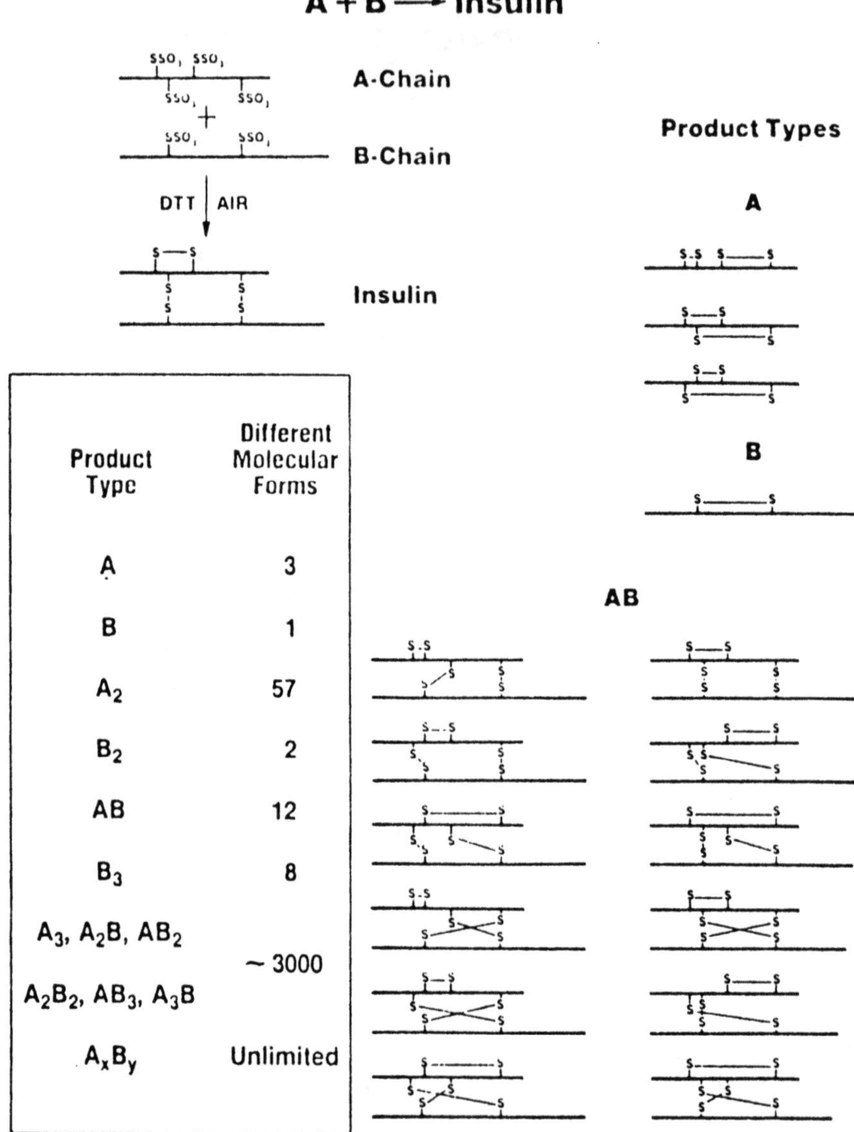

Product Type	Different Molecular Forms
A	3
B	1
A_2	57
B_2	2
AB	12
B_3	8
A_3, A_2B, AB_2 A_2B_2, AB_3, A_3B	~ 3000
A_xB_y	Unlimited

FIGURE 10 Possible sites for disulfide bonds. A rather large number of –S–S– combinations between A and B chains is possible. DTT = dithiothreitol.

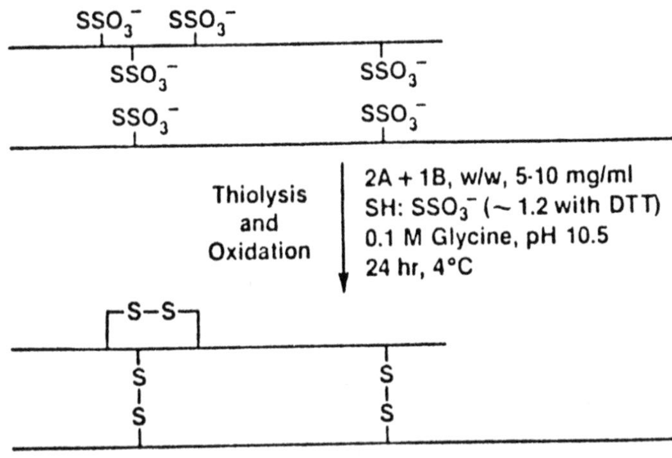

FIGURE 11 Required –S–S– bonds in insulin. Thiolysis of chains and air oxidation step. DTT = dithiothreitol. Courtesy Pierce Chemical Co. (Ref. 10).

to receive the respective end configurations was used to give the final recombinant structure. Insertion of this vector into bacterial cells, for instance, *E. coli*, will give a hormone-producing entity. A dissertation by Olson et al. (12) deals with the problem of the purification of HGH obtained from *E. coli* cloning. Clinical trials of synthetic HGH are underway and considerable success with children has already been achieved.

INTERFERONS

Interferons (IFNs) are proteins secreted by animal cells in response to specific stimuli. The agents which perform this function are called inducers. The contention is that IFNs confer resistance to viral infections.

Presently, there are three types of IFN recognized. They are summarily described by Waldrop (13).

Fibroblast interferon (β-IFN) produced by the cells of the skin, muscle, and connective tissue is prepared by Calbiochem-Behring Co. from human, mouse, and rabbit tissue. HEM Research (Rockville, MD) makes human interferon. *Leukocyte interferon* (α-IFN) is produced from white blood cells, mostly manufactured by Karl Cantell (Helsinki, Finland). *Immune interferon* (γ-IFN) is

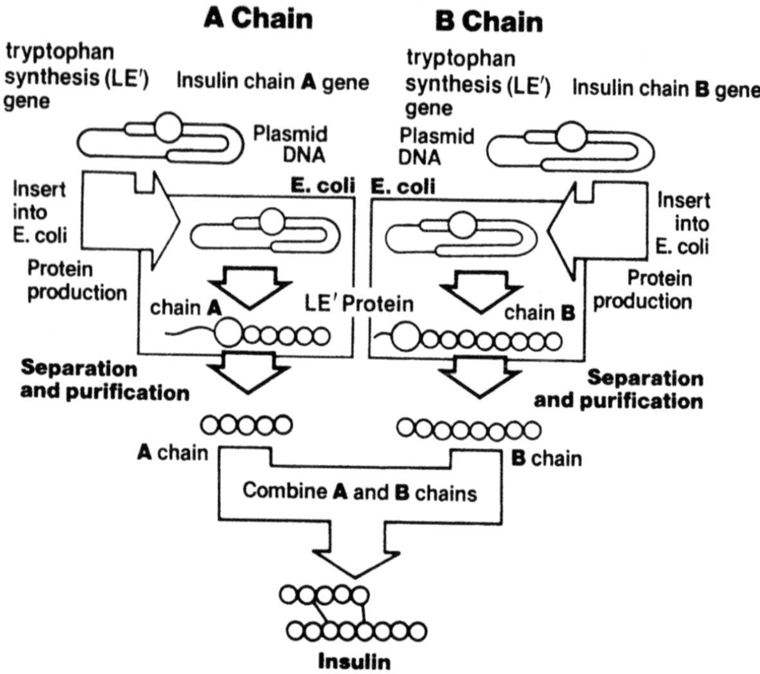

FIGURE 12 Diagram of A–B insulin chain combination. Recombinant plasmids represent a "fixed" protein molecule. LE′: a modified *E. coli* protein. Courtesy Eli Lilly and Co. (Ref. 8).

produced by lymphocytes and fibroblasts. The α and β varieties are also tagged as Type I, while the γ product is Type II.

Potential methods of interferon production were summarized in the Office of Technology publication (Table 3). All three types of interferons are listed with potentials for scale-up, status of development, and inherent problems. Activities are numerous, so in addition to the above manufacturers other news items list California Institute of Technology, Biogen, S.A. (Geneva), Du Pont, Genentech, Hoffman La Roche, G.D. Searle, and Schering-Plough. Most likely there are additional laboratories engaged in interferon research.

The basic approach to the recombinant technique is to infect human white blood cells with a virus to induce interferon production by the cells. Messenger RNA is then extracted by density gradient centrifugation, and analyzed to determine its composition. From the mRNA information it is possible to obtain cDNA with reverse transcriptase. cDNA fractions can then be inserted into bacterial plasmids by recombinant techniques. After plasmid insertion into bacteria (or yeast) the plasmids can be cloned to single colonies. A rather laborious step then ensues in the screening of colonies to isolate interferon-producing bacterial strains, that is, to see if the inserted gene has been expressed in the modified plasmids.

Leukocyte Interferon: IFN*α*

The synthesis of a polypeptide with human leukocyte interferon activity, that is alpha-interferon, was reported in some detail by Nagata et al. (15) in 1980. The procedure is illustrated in Figure 15, and as shown, the scheme follows the above outline, namely via cDNA conversion. It is to be noted that the procedure was repetitive, in that the 500 plasmid group was further subdivided and subjected to several repeat procedures.

Some comments are in order regarding oocyte injection. The foreign DNA entity which is to be injected into eggs to, for instance, assay biologically, must

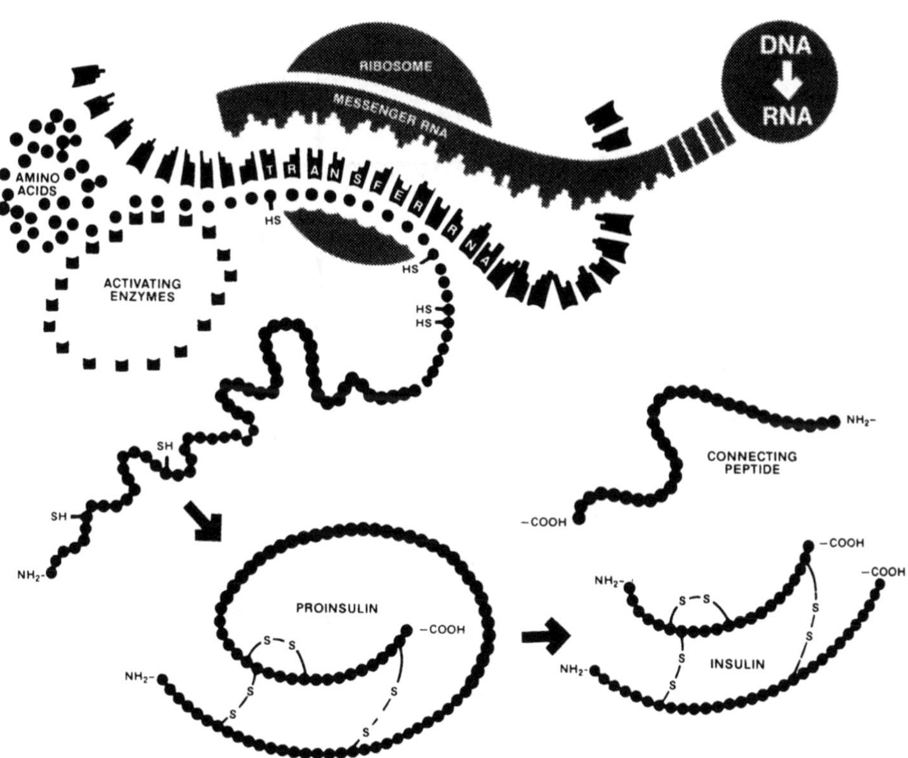

FIGURE 13 Overall synthesis. Pictorial representation of DNA of Insulin protein sequence. Courtesy Eli Lilly & Co. (Ref. 8).

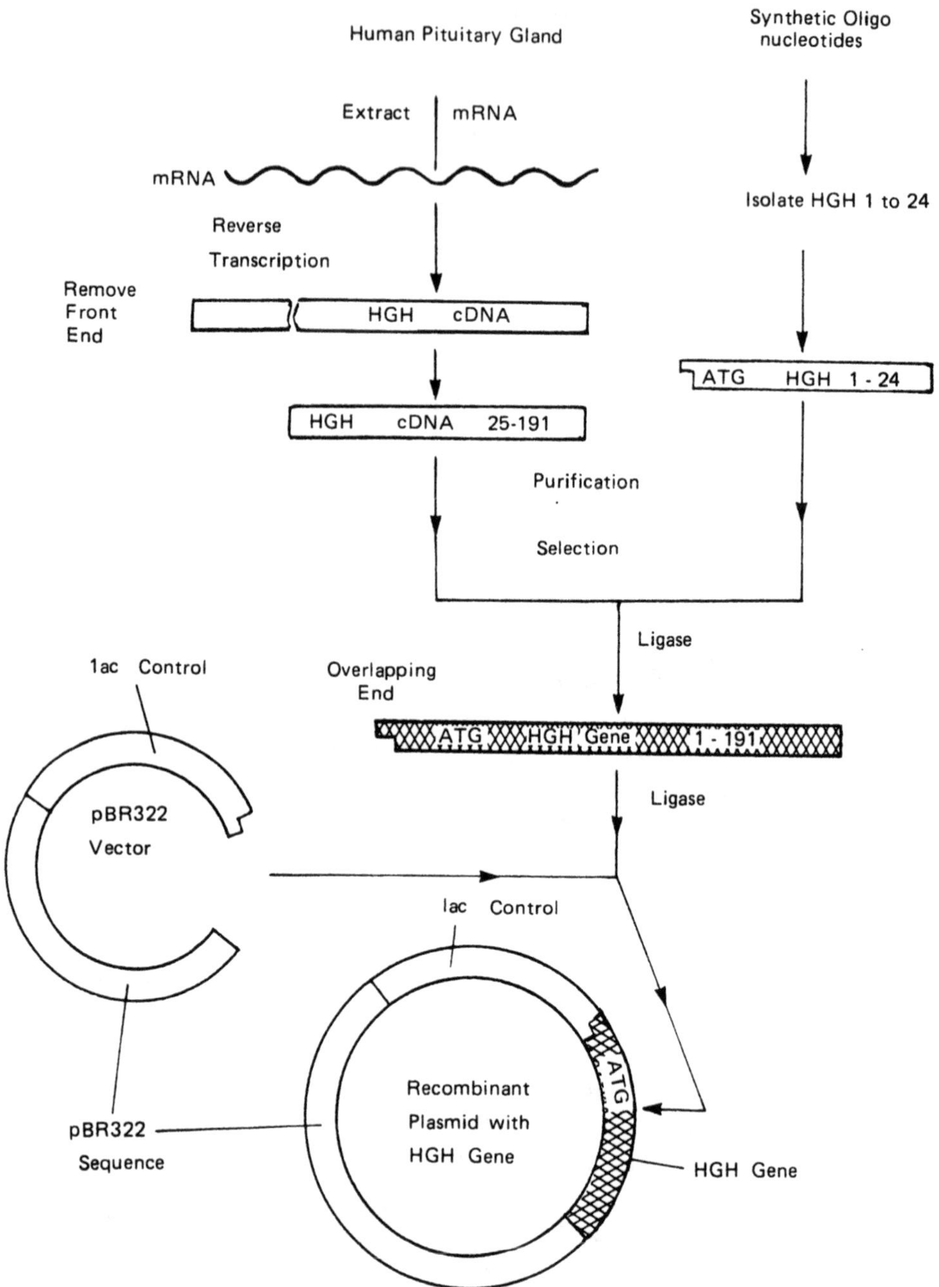

FIGURE 14 Recombinant scheme for human growth hormone. HGH extracted from human pituitary converted to mRNA sequence and reverse transcription to cDNA. Trimming of oversized sequence and synthetic preparation of gene for first 24 amino acids gave complete gene. Insertion into an *E. coli*, pBr 322 plasmid yielded final cloning structure.

TABLE 3 Interferon Production

Means of production	Types of interferon produced	Potential for scale-up	Present projected ($\$/10^6$ units)		Problems	Potential for improvement
"Buffy coat" leukocytes	Leukocyte, 95% Fibroblast, 5%	No	50	–	Lack of scale-up Pathogen contamination	Minimal
Lymphoblastoid cells	Leukocyte, 80% Fibroblast, 20%	Yes	–	≅25	Poor yields Cells derived from tumor	Improved yields Expression of fibroblast interferon
Fibroblasts	Fibroblast	Yes	43–200	≅1–10	Cell culture Economic competition with recombinant DNA	Improved yields Improved cell-culture technology Expression of leukocyte-type interferon
Recombinant DNA	Leukocyte or	Yes	–	≅1–10	Does not produce interferon In vitro stability Poor yields Drug approval Possible economic competition with fibroblast cell production	Improved yields Modified interferons

Source: Office of Technology Assessment.
Courtesy Westview Press (Ref. 14).

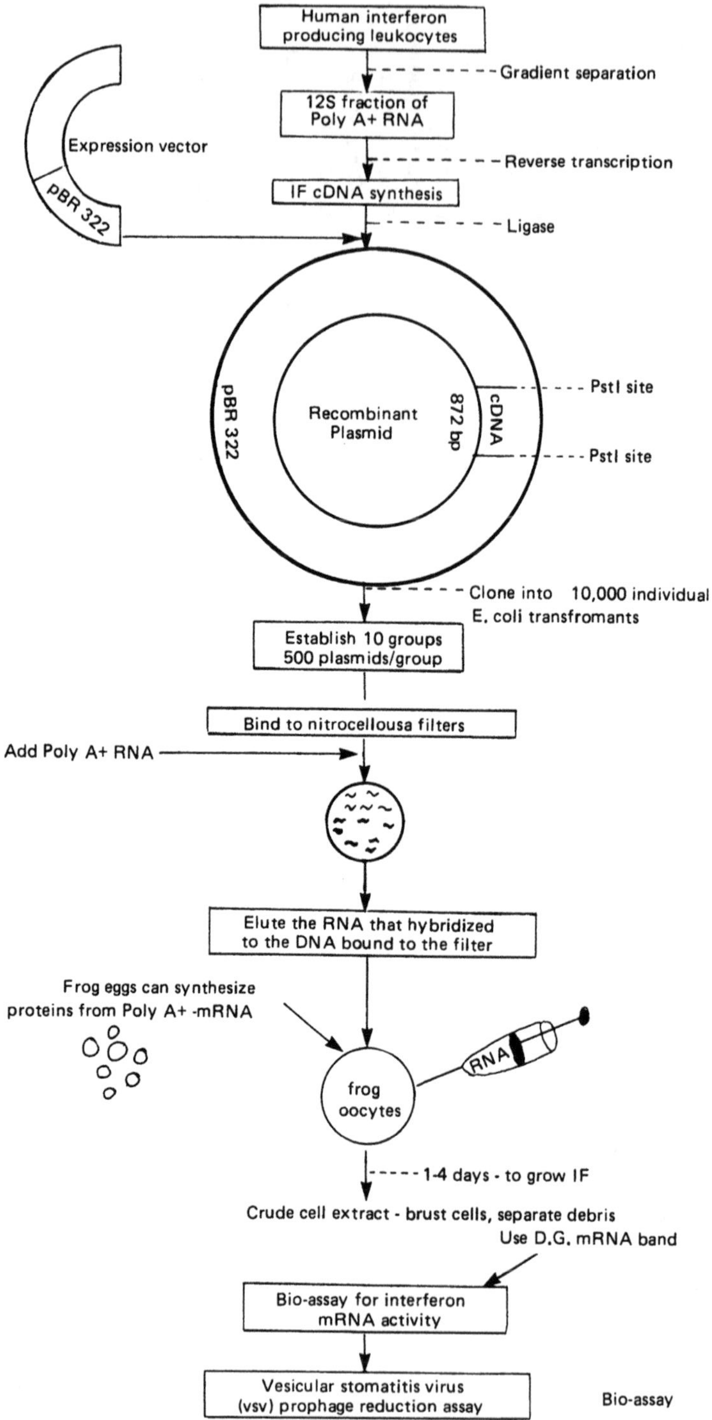

Human interferon
producing leukocytes
Gradient separation
12S fraction of
Poly A+ RNA
Expression vector
pBR 322
Reverse transcription
IF cDNA synthesis
Ligase
pBR 322
Recombinant
Plasmid
872 bp
cDNA
PstI site
PstI site
Clone into 10,000 individual
E. coli transfromants
Establish 10 groups
500 plasmids/group
Bind to nitrocellousa filters
Add Poly A+ RNA
Elute the RNA that hybridized
to the DNA bound to the filter
Frog eggs can synthesize
proteins from Poly A+ -mRNA
frog
oocytes
RNA
1-4 days - to grow IF
Crude cell extract - brust cells, separate debris
Use D.G. mRNA band
Bio-assay for interferon
mRNA activity
Vesicular stomatitis virus
(vsv) prophage reduction assay
Bio-assay

be injected into the nucleus. This is a protective move to shield the DNA template from possible attack by nucleases present in the cell cytoplasm. Two basic injection techniques are in use. In one method the micropipette, usually a glass tube, is aimed for the center of the oocyte where the nucleus should be located (16). The other method attempts to displace the nucleus and bring it to the cell wall surface by gentle centrifugation (17). This makes the nucleus more accessible to injection.

The biological assay is analogous to the use of a "blank" determination in analytical chemistry. Two sets of oocytes, usually live frog eggs, are infected with the same amount of a virus, specifically vesicular stomatitis virus plaque. One set receives an injection of interferon extract. The difference in virus titer is an indication of the effectiveness of interferon to retard virus multiplication.

A similar procedure was reported by Edge and co-workers (18) in 1981. The basic procedure is much like that of Nagata et al., but some variations are of sufficient interest so that a condensed outline is indicated. Edge et al. describe the synthesis of IFN-α DNA fragments, based in part on the published work of Weissman's group (19). A 514 bp fragment of ds-DNA was synthesized in toto. The DNA sequence coded for IFN-α_1 (166 amino acids). It contained initiation and termination signals, as well as proper restriction enzyme sites to effect plasmid insertion. The parent plasmid was pAT 153 (20) which was cleaved with *Eco*RI and *Bam*HI, followed by 3′-oligonucleotide extensions. Intermediate steps comprised insertion of the *E. coli* lactose operon and a part of the β-galactosidase gene (obtained from plasmid pPM18). The synthetic interferon gene was then introduced into the modified plasmid intermediate; therefore, the final recombinant plasmid contained the *Eco*RI site, *lac* PO (promoter/operator), *Bam*HI site, IFN gene, and *Sal*I, in clockwise order. Ampicillin resistance from the original plasmid was retained, while tetracycline resistance was destroyed by *Bam* and *Sal* cleavage.

The synthesis required the preparation of 66 oligodeoxyribonucleotides varying in length from 14 to 21 units and one deoxydecanucleotide. Synthesis was carried out by solid-phase reaction followed by enzymatic ligation of the fragments.

An interesting variation of the genetic code table is shown in Table 4, which was developed by the investigators primarily for experimental synthesis. The significance of the numbers in the table is as follows:

FIGURE 15 Scheme for producing leukocyte interferon (15). Sequential steps involve infection to generate interferon, separation by density gradient procedure, introduction to an *E. coli* plasmid, transmission through a nitrocellulase technique, addition of poly (A), elution and finally bioassay.

TABLE 4 Variation of the Genetic Code Table

	U		C		A		G		
U	Phe	3(4)	Ser	3(4)	Tyr	3(1)	Cys	3(4)	U
	Phe	5(4)	Ser	4(4)	Tyr	1(3)	Cys	2(1)	C
	Leu	2(2)	Ser	2(2)	Ochre	1(1)	Opal	–	A
	Leu	0(4)	Ser	0(0)	Amber	–	Trp	2(2)	G
C	Leu	1(0)	Pro	0(3)	His	2(2)	Arg	6(0)	U
	Leu	3(8)	Pro	1(2)	His	1(1)	Arg	2(0)	C
	Leu	2(1)	Pro	0(1)	Gln	7(2)	Arg	0(1)	A
	Leu	14(7)	Pro	5(0)	Gln	3(8)	Arg	0(0)	G
A	Ile	0(0)	Thr	3(2)	Asn	2(2)	Ser	0(0)	U
	Ile	7(7)	Thr	6(4)	Asn	4(4)	Ser	4(3)	C
	Ile	0(0)	Thr	0(3)	Lys	4(4)	Arg	3(6)	A
	Met	6(6)	Thr	0(0)	Lys	4(4)	Arg	1(5)	G
G	Val	2(1)	Ala	6(4)	Asp	4(5)	Gly	1(0)	U
	Val	1(2)	Ala	1(3)	Asp	7(6)	Gly	2(1)	C
	Val	2(0)	Ala	2(2)	Glu	7(6)	Gly	0(2)	A
	Val	1(3)	Ala	1(1)	Glu	8(9)	Gly	0(0)	G

Reprinted by permission from *Nature* 292 (5824): 757 (1981) (Ref. 18).

Codon use for *natural* sequence. The *numbers* in *parentheses*; assume that there are 8 Phe in the chain (upper left in table). The *4* will be coded by UUU and *4* more by UUC.

Codon use in *synthesis, open numbers*. The investigators found that 3 Phe can be coded by UUU and 5 Phe by UUC.

Interferon seems to wear different cloaks at different times. It is tempting and vexing, appearing in 20 or more molecular versions. It is considered to be *species specific*, a one-on-one situation where a certain type is beneficially active in what one might call a covalent situation, but inactive or possibly deleterious in a nonrelated situation. An overview of presently known behavior is nicely presented by Bodde (21), which leaves the question: is interferon a bonanza or a bomb? Nevertheless, its proven benefits in antiviral activity, in the slowing of cell division and retardation of tumor growth, the latter of tremendous interest to cancer research, are more than sufficient reasons to emphatically pursue its manufacture. A text on properties and uses of interferons was published by Friedman (22). Also, a delightful, narrative account of research experiences by Weissmann (23) is worthwhile reading. It bears the intriguing title *The Cloning of Interferon and Other Mistakes.*

An impressive article dealing with interferon inducers by Hoffman (23a) throws some light on the seemingly erratic results that have been reported in a number of studies. Hoffman reasons that interferon behavior may have to be interpreted on the basis of symbiotic interaction between interferon and protein-aceious-inducing agents. It is pointed out that the induction phenomena are difficult to differentiate from other cytokinetic effects such as virostatic, cancerostatic and immunomodifying effects. A diagram is presented which attempts to schematize the essential interrelated processes associated with interferon action, including presumable feedback phenomena and a mediation role of double-stranded RNA.

IFN production and extraction from cells calls for protein as well as inducer removal. If recombinant techniques are used, the native bacterial protein must be eliminated. The purification problems require a major effort in the production of injectable IFN. The problems and effective procedures are presented by Fulton (24) of the Amicon Corporation.

Immune Interferon: IFN-γ

A very detailed account of the production of a recombinant structure to produce *gamma* interferon is given by Gray et al. (25). The creation of a recombinant plasmid was part of a study to identify a cDNA sequence in a cDNA library. The preparation of the plasmid in a condensed outline scheme is as follows:

A sequence of 1250 bp *Pst*I insert was obtained from plasmid p69. Enzyme digestion, followed by gel electrophoresis identified a desired 1100 bp fragment which was eluted. Synthetically prepared deoxyoligonucletodies were phosphorylated and attached to the fragment and a 115 bp fragment was then separated by electrophoresis and recovered by elution: containing *Eco*RI, *Pst*I, and IFN-γ DNA.

Plasmid pLeIFA *trp* 103 was cleaved to remove the LeIFA sequence. A vector fragment of about 3900 bp was recovered by gel separation. This fragment and the above 1115 bp fragment were ligated to yield a recombinant plasmid containing the IFN-γ gene sequence. This plasmid was introduced into *E. coli* K-12 strain 294 to yield tetracycline-resistant colonies. The polypeptide produced by expression of the inserted DNA sequence in *E. coli* had the properties of authentic human IFN-γ. Bioassay was conducted in oocytes of *Xenopus laevis*, a primitive African toad.

Ongoing research reported by Gray and Goeddel (26) describes findings on the DNA sequence of the human IFN-γ gene. The structure was found to have three introns (repetitive sequences). The conclusion is reached that there is only one gene for IFN-γ. The observation in other studies that IFN-γ may appear as two separate components is possibly due to posttranslational processing of the protein. The restriction enconuclease map and the sequence of the gene region are reported in detail.

Human Beta-Interferon: IFN-β

A procedure for producing a clonable beta-IFN entity was described by Reyes et al. (27). As one would expect the sequence of steps is very much like those used for IFN-α and IFN-γ. So, the coding portion of the IFN-β gene was inserted after the promoter sequence and the RNA start site of the thymidine kinase (TK) gene. The TK gene was derived from a herpes simplex virus type 1. The chimeric plasmid thus constructed resulted in expression of IFN activity when microinjected in nuclei of oocytes of *Xenopus* (see IFN-γ) and also when transfected into mouse cells followed by superinfection with herpes simplex virus.

SUPERBUG

A conceptual scheme for the creation of a "superbug," that is, a bacterium capable of metabolizing hydrocarbons, was put forth by Hopwood (28). It suggests the formation of plasmids with gene inserts from four different bacterial strains of *Pseudomonas putida*. Thus, camphor and octane scavenging plasmids are created in one operation, and xylene and naphthalene genes are introduced into two other plasmids. Crossing and plasmid recombination leads to two entities. Subsequent further screening should yield the final plasmid capable of metabolizing all four hydrocarbons. The schematic is shown in Figure 16.

Interestingly, the genes for metabolizing the hydrocarbons are not located in the chromosomes, but in plasmids contained in various *P. putida* strains. The genetic action arises from enzymes which are coded by the respective DNAs. As shown in the diagram the CAM and OCT plasmids can be combined into a single plasmid entity. The XYL and NAP plasmid will exist as separate units in another plasmid. In the crossing process one type of plasmid will transfer to another one, so that the final plasmid structure may acquire all of the gene properties inherent in the initially used strains.

CREATION OF OTHER SYNTHETIC RECOMBINANTS

As pointed out earlier, the general scheme of preparing vectors for insertion of natural or modified DNAs is rather simple conceptually, albeit often complicated experimentally. Also, the various procedures are repetitive to an extent, so that one encounters enzyme cleaving of plasmid or viral DNA, tailoring of foreign DNA, either to trim and/or poly tail, insertion into vector with vector

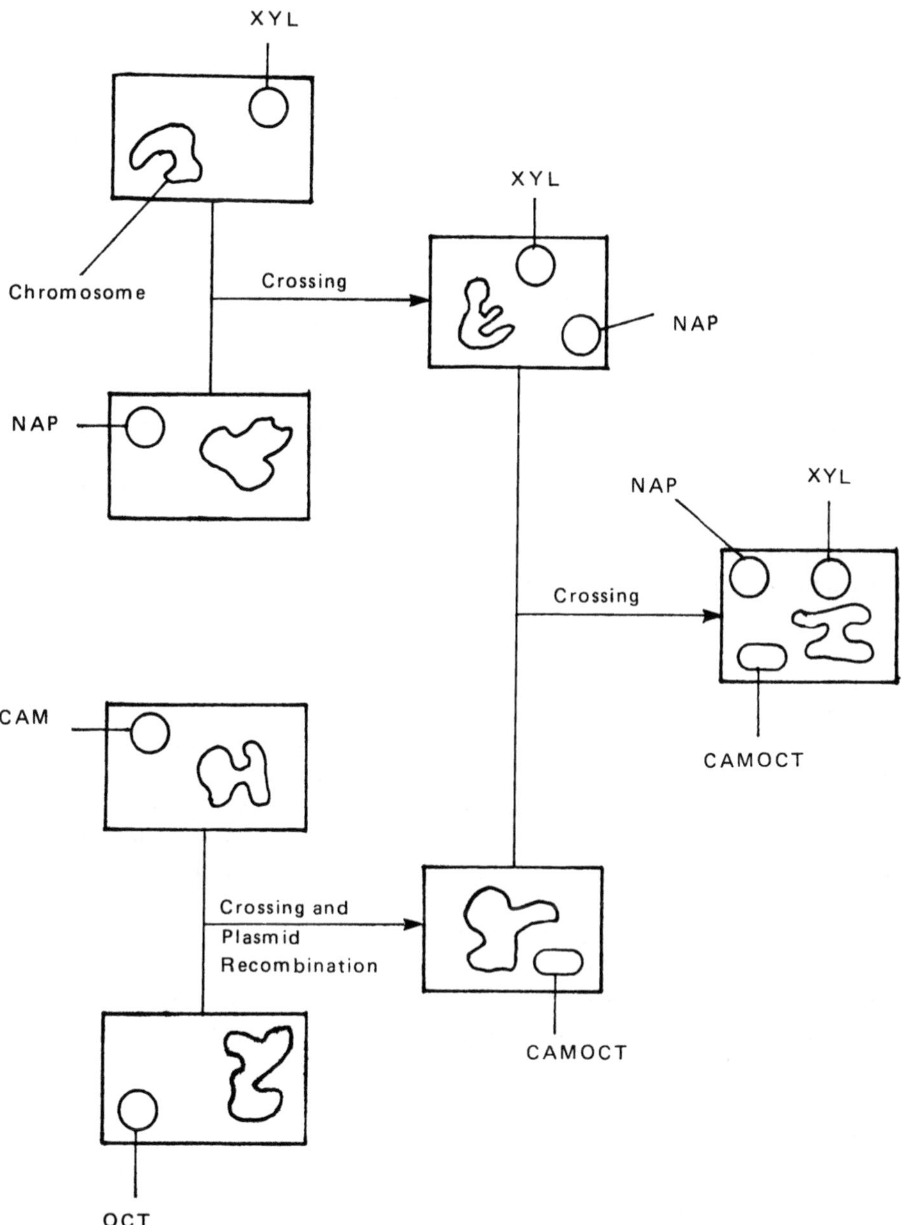

FIGURE 16 Structure of superbug. Insertion of *Pseudomonas putida* genes create plasmids capable of metabolizing hydrocarbons. CAM - camphor; OCT - octane; XYL - xylene and toluene; NAP - naphthalene.

transformation into host and cloning. The diagram visualizing these steps begin to look alike, as indeed they are. Thus, it should be acceptable to describe some additional creations of recombinant entities without resort to diagrammatic visualization. If any procedure thus described is of more than passing interest to the reader, the original publication should be consulted. There, any unusual recombinant steps will generally be illustrated in detail.

ANTIHEMOPHILIC AGENTS

A British company, Speywood Laboratories Ltd, is tooling up for increased production of blood proteins essential to the treatment of hemophilia, in particular antihemophilic factor VIII (a modifier protein). Genetic engineering techniques using yeast as a vector, are counted upon for quantity production of the proteins to overcome the present problem of obtaining only very small quantities of high purity from pig and human blood (29).

The molecular cloning of the gene for human antihemophilic factor IX was described by Choo et al. (30). A deficiency of this factor is responsible for Christmas disease or hemophilia B. The isolation and partial characterization of a λ-recombinant phage which contains the gene is detailed.

SYNTHETIC VACCINES

The production of conventional vaccines from virus-infected mammalian tissues is, at present, still the only source of supply. The presence of irrelevant microbial antigens, of proteins and other constituents in vaccines produced in this manner can be a serious drawback. These "foreign" matters can result in unwanted and potentially dangerous reactions in the mammalian body. Consequently, the production of synthetic vaccines of the highest purity is a most attractive area for recombinant techniques.

A tabulation of medical and therapeutic materials, Table 5, was presented by Sherwood and Atkinson (31) and it is significant that 4 out of 15 items are of the vaccine type. Efforts presently under way deal with vaccines against infectious diarrhea in pigs and calves, and numerous mammalian viral infections, as, for instance, diphtheria, influenza, etc.

Hepatitis B Vaccine

An overview of synthetic vaccine research by Zuckerman (32) discusses the work of Dreesman et al (33) and the studies of a number of other investigators. Dreesman's group prepared a vaccine against hepatitis B infection by synthesizing two peptides containing a hydrophilic region identified by computer analysis

TABLE 5 Medical and Veterinary Therapeutic Materials Cloned and Expressed in *E. coli* (December 1980)

By direct gene cloning:
Hepatitis B core antigen
Hepatitis B surface antigen
From cDNA by reverse transcription from mRNA:
Rat preproinsulin
Rat proinsulin
Rat growth hormone
Human leukocyte interferons
Human fibroblast interferons
Fowl plague virus surface antigen
Foot and mouth virus surface antigen
By chemical synthesis of the gene:
Human somatostatin
Human growth hormone[a]
Human insulin A chain
Human insulin B chain
Human proinsulin[a]
Human thymosin α-I hormone

[a]By a combination of chemical synthesis and reverse transcription of cDNA from mRNA.
Courtesy *Chemistry and Industry* (Ref. 31).

of a hepatitis B surface antigen. The peptides elicited the production of antihepatitis B surface antigen antibody in mice with only a single injection. It seems entirely feasible that vaccines in general could be synthesized as peptide fragments containing the active sequence. Thus the ever present problem of side reactions and possible toxicity from normal vaccine production might be eliminated.

Edman et al. (34) describe the creation of a recombinant plasmid which can be cloned in *E. coli* to synthesize hepatitis B surface and core antigen gene sequence (HBcAg). The preparation of the plasmid is shown diagrammatically in considerable detail. A tryptophan (*trp*) operon regulatory region from an *E. coli* plasmid was utilized as a control factor. A plasmid labeled NBVDNA, containing the core Ag gene, was digested to isolate that gene, which then was introduced into the plasmid which carried the *trp* promoter-operator. The result, a plasmid labeled pCA 246, produced the highest levels of HBcAg, compared to several other structures.

Hepatitis B Virus Surface Antigen [HBsAg (35)]

An HBsAg coding sequence (obtained as an 835 bp fragment) was joined to the yeast alcohol dehydrogenase promoter (ADHI) and introduced to a pBR322 plasmid. The sAg gene was excised from a pHVB-3200 plasmid and inserted into the modified pBR 322. The resultant structure then contained sAg gene and ADHI. Culturing led to the production of particles of ~22 nm size, similar to those produced in the natural infection with hepatitis B virus. The immunogenicity of the particles in animal experiments was as high as that of the natural particles.

Synthetic Peptide against Foot-and-Mouth Disease (FMD)

FMD is an animal disease of major importance. It is prevalent in domestic and wild ruminants and also presents a virulent infection in swine. In Australia and North America the disease is essentially nonexistent because of very stringent controls. However, its incidence in the rest of the world is still alarmingly high. It is estimated that more than one billion doses of vaccine are needed each year (36). The presently used vaccines are produced by inactivation of virus harvests obtained from animal tissues and so are subject to the usual problems of vaccines, such as stability and toxicity.

Consequently, the production of a synthetic vaccine is a very attractive proposition. It is not surprising that considerable activity is underway on a worldwide scale. The most recent publication by Bittle et al (36) describes the chemical synthesis of peptides that were effective antisera against FMD virus. Similarly, Boothroyd et al. (37) were successful in constructing recombinant plasmids containing cDNA copies of the viral RNA and established sequences which agree with published compositions. As Bittle's work became known, Küpper's group (38) published their results in the expression of a FMD virus in *E. coli*.

Küpper's procedure involved modifying a pBR322 plasmid by inserting cDNA from FMDV (V-virus), thus producing a donor plasmid containing 5500 bp. A separately constructed vector plasmid pPLc24 was then ligated to the donor plasmid. The final product, labeled pPLVPl (4080 bp), contained Amp^r, FMDV insert, a sequence for MS2 polymerase and single restriction sites for *Eco*RI, *Bam*HI, *Hind*III, and *Hae*II. This plasmid was found to transform into *E. coli* K12 and produce a viral capsid protein sequence designated VPl. Antigenic activity was obtained with the pPLVPl plasmid but not with its two precursors. Küpper's publication contains the structural diagrams of the plasmids and information on a FMDV genome map. Also of interest is an earlier report by Kimber et al. (39) where cloning in *Bacillus subtilis* was successful.

NONRECOMBINANT GENETIC ENGINEERING

Monoclonal Antibodies

Monoclonal antibodies (MAb) are genetically engineered creations without the use of recombinant techniques. The basis for their production is the science of immunology. To ensure a clear understanding of the procedure used to grow MAbs it is necessary to briefly cover some basic immunologic concepts.

Antigen

An antigen is basically any foreign protein to which an organism can mount an immune response. In nature, antigens are usually an infectious agent such as a fungus, a virus, or a bacterium. The presence of the foreign antigen triggers the immune system to respond so as to neutralize, or eliminate, the foreign invader.

Antibody

The immune response results in the formation of sensitized lymphocytes, the production of antibody, or both. Antibody activity is found in the gamma-globulin fraction of the blood serum. In general, a given antibody is very specific for a given antigen. The antibody reacts with the antigen on a "lock and key" basis, such that the specific antibody reactive site fits three dimensionally into a specific protein configuration on the antigen called the antigenic determinant.

Lymphocytes

Lymphocytes are the specific cells in the body responsible for carrying out the immunologic response. There are two major types of lymphocytes called T cells and B cells, which are found predominately in the lymph nodes and spleen of the individual. Lymphocytes also migrate in the blood stream, and 20–40% of the white blood cells are lymphocytes. T lymphocytes pass from the bone marrow through the thymus on the way to the lymph nodes during fetal development. These T cells are responsible for cell-mediated immune reactions. B lymphocytes also arise in the bone marrow and migrate directly to the lymph nodes. Upon exposure to antigen, B cells undergo transformation into plasma cells which are responsible for antibody or immunoglobulin production. The formation of the lymphocytes is sketched in a simplified manner in Figure 17.

Immunoglobulins (Ig)

Immunoglobulins are proteins of considerable variability in their composition. Organizationally there are five classes of immunoglobulins in humans: IgM, IgG, IgA, IgD, and IgE. Using this nomenclature Ig stands for immunoglobulin and the letters M, G, A, D, and E are alphabetical codes which developed historically. Certain properties and characteristics of the five immunoglobulin classes are shown in Table 6. Each of the five classes of immunoglobulin have different

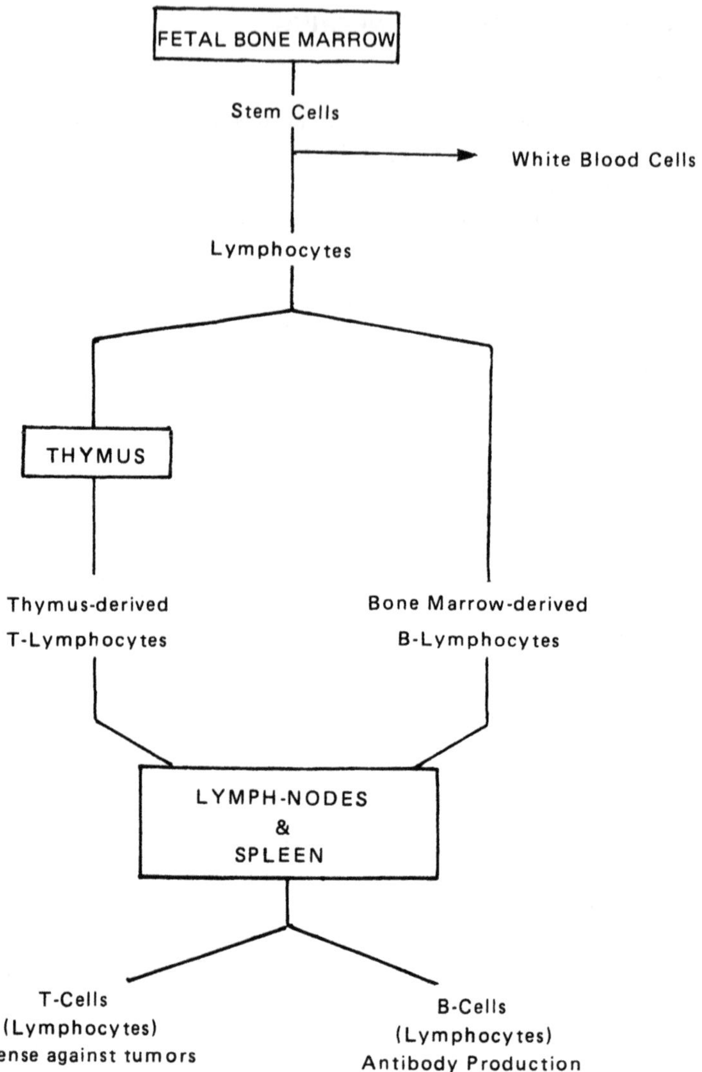

FIGURE 17 Formation of lymphocytes. Originating in fetal bone marrow T-cell lymphocytes and B-cell lymphocytes follow different paths of maturation.

TABLE 6 Properties of Immunoglobulin Classes

	IgG	IgA	IgM	IgD	IgE
Molecular weight	150,000	170,000 or polymer	900,000 (5 X 180,000)	160,000	200,000
Mean half life (days)	23	6	5	3	2.5
Normal serum concentration range in adult (mg/100 ml)	800–1600	150–400	50–200	0.3–30	0.0015–0.2
% Total immuno-globulins	80	13	6	1	0.002
Relative rate of synthesis	4+	2+	2+	1+	Unknown
Relative rate of antibody activity	3+	3+	3+	1+	3+
Relative heat sensitivity	Stable	1+	2+	4+	3+
Number of basic 4-peptide units	1	1 in serum 2 in secretion	5	1	1
Complement fixation	Yes	No	Yes	No	No
Placental transfer	Yes	No	Yes	No	No
Fixation to mast cells and basophils	No	No	No	No	Yes

biologic properties and functions. Upon antigen exposure the first immunoglobulin response is that of IgM formation, which is followed by a much larger peak of IgG formation. IgG can cross the placental membrane and therefore the newborn infant is protected by maternal IgG. Both IgM and IgG have the ability to fix complement and form a major defense against microbial invasion. IgA is mainly a secretory antibody, IgD appears to be primarily a lymphocyte surface receptor. IgE is triggered in response to parasitic infections and also binds to skin and membrane surfaces, thus being responsible for inhalant allergy such as asthma and hayfever.

Structural Characteristics of Immunoglobulins

All immunoglobulins share a characteristic structure which is in the form of a *Y* and clearly shows up in electron photomicrographs (40). The basic immunoglobulin structure is illustrated in Figures 18a and b (41). Each immunoglobulin unit is made up of four protein chains, two heavy chains and two light chains. The so-called "constant region" of the four chains is basically the same in all antibodies of a given class, and this constant region determines the biologic properties of that given class of immunoglobulin. The "variable region" of the four chains in each immunoglobulin molecule differs from antibody to antibody and allows the antibody to recognize and bind to a specific antigen. The "hypervariable regions" at the end of the antibody molecule are able to form various spatial configurations which result in highly specific antigen-combining sites and allow for the "lock and key fit" with the antigen.

Monoclonal Formation

The production of antibodies (42) is illustrated in Figure 19. At the left the natural formation of antibodies is presented. Stimulation of the immune system by antigen injection results in antibody production from lymphocytes. The antibodies that are formed in response to the various antigenic determinants are a heterogeneous mixture with multiple specificities. B lymphocytes in the lymph nodes and spleen undergo transformation into plasma cells which produce the antibodies. From there the antibodies diffuse into the blood stream to interact with the foreign antigen. Once an organism has been exposed to a foreign antigen, a small number of B lymphocytes, which will recognize that antigen in the future, are stored in the lymph nodes and spleen as "memory cells."

The right lane of the figure depicts the process of monoclonal antibody formation. Lymphocytes are removed from the spleen and fused with melanoma cells to form a *hybridoma* complex. Melanoma cells are lymphocytes which have become malignant (cancerous). The hybrid cells, now benign, can then be cloned so that only pure monoclonal antibody is produced which is specific for one antigenic determinant. An interesting and often quoted citation from Milstein (43) is as follows "fuse a normal lymphocyte or plasma cell with a myeloma cell and thus immortalize the expression of the plasma cell's antibody secretion."

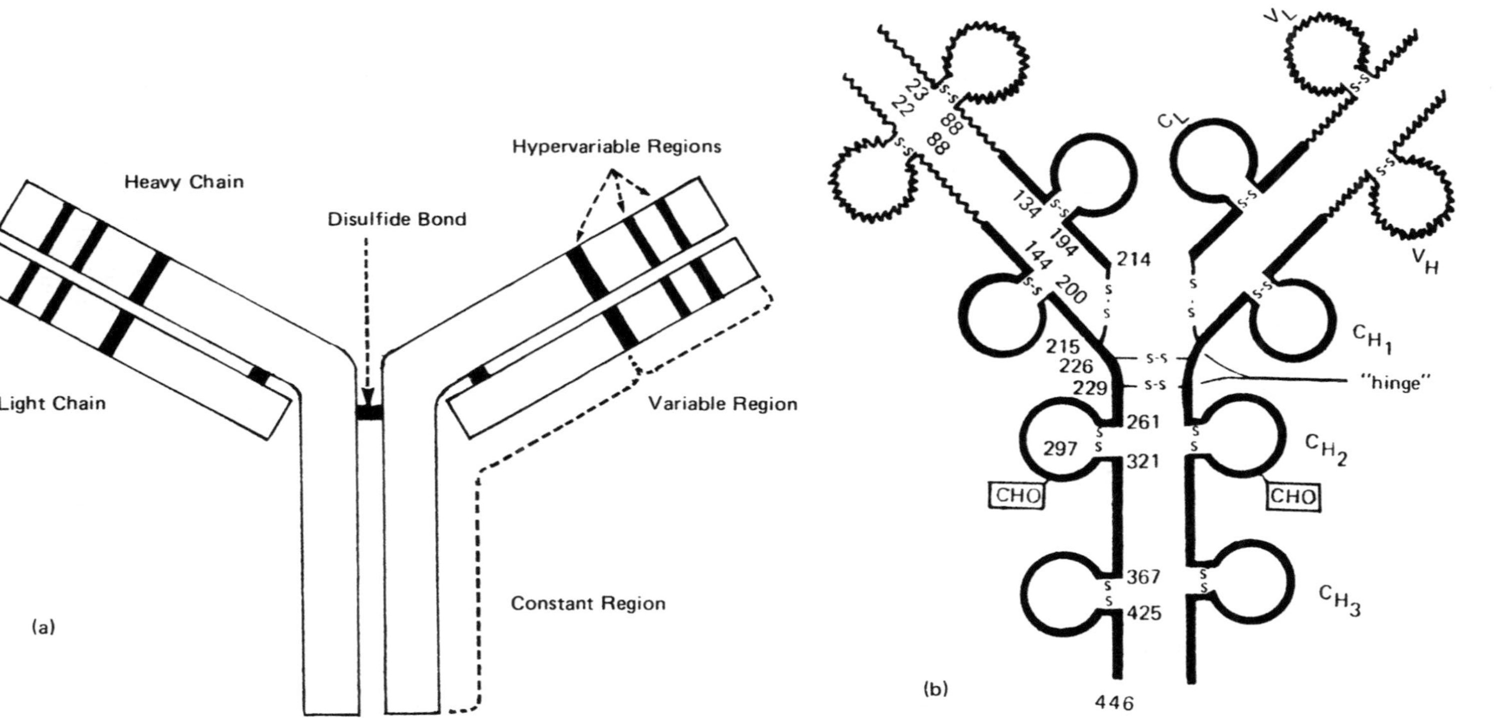

FIGURE 18 (a) Structure of antibody. Idealized diagram showing constant and variable regions. Note the concept of heavy and light chains. (b) Antibody structure. Details of amino acid distribution. Linear periodicity in amino acid sequences suggests that light (L) and heavy (H) chains have repeating domains, each with about 110 amino acid residues and an approximately 60-membered S-S bonded loop. Domains with variable sequences are represented by jagged lines (V_L, V_H); those with invariant sequences in a given class of H or L type are represented by smooth lines (C_L, C_{H1}, C_{H2}, and C_{H3}. Numbered positions refer to cysteinyl residues that form S–S bonds or to the point of attachment of an oligosaccharide (CHO). Courtesy Harper & Row (Ref. 41b).

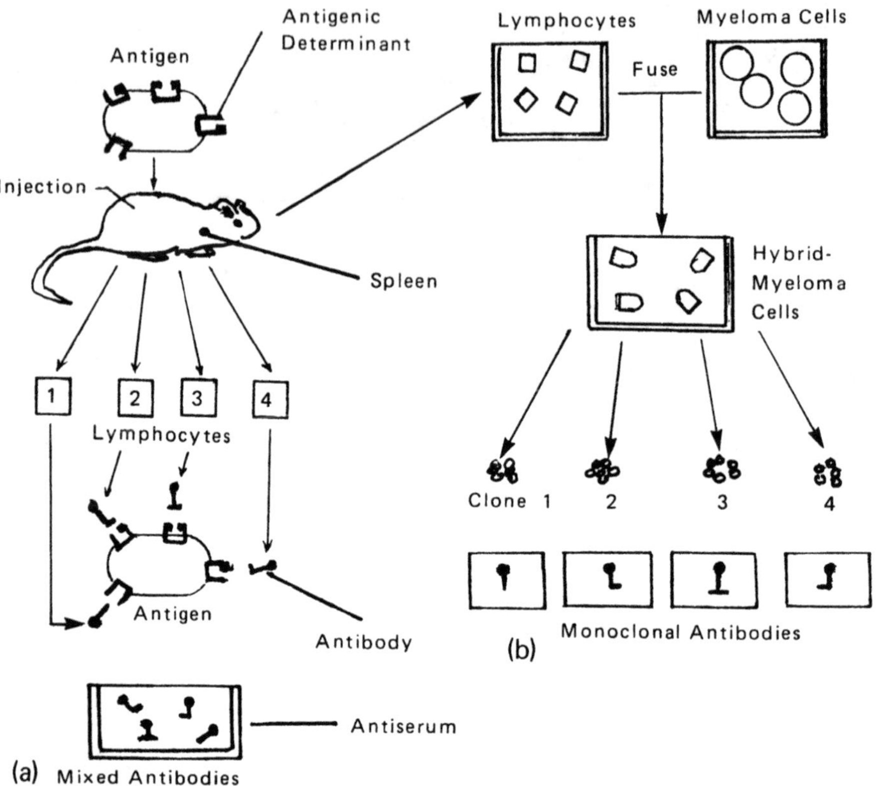

FIGURE 19 Monoclonal antibodies. (a) Formation of natural antibodies as a mixture through injection of an antigen into mouse. (b) Hybridoma formation and culturing of pure monoclonals.

The "fusing" step is carried out in a manner similar to that described under somatic cell fusion in plants (Chap. 9). A suitable liquid medium for cell fusion is polyethylene glycol in a concentration of about 40%. In addition, a so-called HAT medium is used in a subsequent screening step to isolate the desired clone type. The HAT medium is a very important adjunct in cell growth and cell screening procedures. It contains hypoxanthine, aminopterin, and thymidine. The medium is used with mammalian cells which can incorporate exogenous hypoxanthine and thymidine and so are able to bypass the aminopterin-blocks in the purine and thymidylate metabolism. Mutant cells which lack the enzymes hypoxanthine phosphoribosyl transferase (HGPRT) or thymidine kinase (TK) will not grow in a HAT medium. However, when a TK^- cell is fused with a $HGPRT^-$ cell, a hybrid cell is formed which will grow in the HAT medium. A further use of HAT medium is the selection of transformants of TK^- cells which have been treated with DNA that contains the thymidine kinase gene derived from herpes simplex virus.

APPLICATIONS OF MONOCLONAL ANTIBODIES

Monoclonal antibodies are available commercially from a number of sources. The literature on monoclonals is growing rapidly and it would be redundant to itemize all the references available. It should be acceptable to mention just a limited number of applications, either already in use or as suggested possibilities.

From a literature aspect, a Special Hybridoma Issue of the *Journal of Immunological Methods* (44) would be worthy of attention. There, at least 12 pertinent publications deal with a variety of subjects such as enzyme-linked immunosorbent assay for screening MAbs, development of a cell line secreting MAbs, derivation and characterization of a MAb hybridoma antibody specific for human alpha-fetoprotein (AFP), production of MAbs in serum-free media, etc.

A most important use of MAbs is an analytical one in identifying cell type. Thus, Potter of the National Cancer Institute is quoted as:

> Monoclonal antibodies are to immu-
> nology what the electron microscope
> is to anatomy, *and* they are the tool
> that allows us to study so many
> things at the chemical level.

Cell type recognition is described by Barnstable (46). Lee and Kaufman (47) report on the use of MAbs in the analysis of myoblast development. Also Milstein and co-workers (43) describe analytical uses of MAbs.

Monoclonal Antibody Screening

The use of *enzyme-linked immunosorbent assays* (ELISA) has become practical with the availability of monoclonal antibodies. A New England Nuclear bulletin (48) describes the ELISA technique in the diagnosis of infectious diseases in human and veterinary medicine. Circulating antibodies in an infected host can be detected by using enzyme-linked anti-immunoglobulin.

The diagrams in Figure 20 illustrate schematically the procedure. Thus, the organism-specific antigen is immobilized on a solid surface, such as a test tube wall. Incubation of the sample (serum, culture) containing the infectious agent with the antigen on the solid support results in the attachment of the appropriate antibody to the antigens. The serum or culture medium is then washed

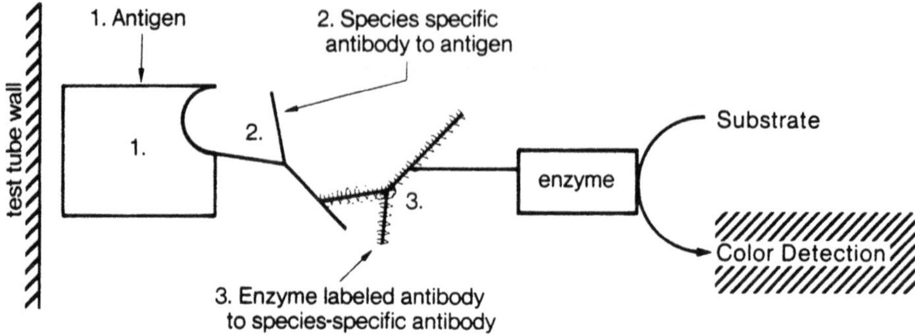

FIGURE 20 ELISA technique. ELISA is Enzyme-linked immunosorbent assay which uses monoclonals for identification of antibodies. Courtesy New England Nuclear (Ref. 48).

from the solid surface while the antibody remains bound to the antigen. Then, an enzyme-labeled antibody specific to the species-specific immunoglobin is added. After incubation, the enzyme-labeled anti-immunoglobin solution is decanted and an enzyme substrate solution is added to the solid support phase. If specific antibodies to the antigens of the infectious organism are present, a color reaction takes place.

Much work is underway to establish if MAbs are effective as antitumor agents. Reports of such investigations have been made by Hellström et al. (49), Bernstein et al. (50), Blythman et al. (51), and Peng et al. (52). In a somewhat related situation, Lerner (53) reports MAbs have been found helpful in reversing rejection of a transplanted kidney.

FACT OR FICTION?

Already closer to fact than fiction are some of the ideas which are being bandied around for the use of "engineered" molecules and bacteria. Some wayout applications and some startling accomplishments are conceptualized in a challenging article by Angier (54). The *biochip* consisting of engineered proteins will replace the silicon microchip. The action of the biochip is described as "organic molecules. . . that dance at the touch of an electric current, winding or unwinding, passing hydrogen atoms from one end to the other, kinking up or straightening out." Besides greatly increased speed of operation, the devices would pack an enormous amount of memory into each chip. The statement is made that such an organic computer might be one cubic centimeter in size and possess 10 million times the capacity of today's instruments. The "engineered" molecular motions are the means for information transmission. The unique quality of proteins to conduct self-assembly is envisioned as a potential ability to have the components undergo self-assembly. Finally, research is already underway to adapt biochips to give sight to blind people.

The subject also was treated by Robinson (54a) in a scientific manner. Carter of the U.S. Naval research Laboratory is quoted: "If we had more support, we could have more results." Robinson's article is a report on a workshop dealing with molecular electronic devices.

GENETIC ENGINEERING EXTENDED

Mining through the use of bacteria (microbiological mining) is becoming an essential means of recovery in the minerals industry. While one might consider this application to be a fringe area of genetic engineering, it would seem that it belongs as an extension in a useful utilization of bacterial capabilities. Obviously, one may expect the development of recombinant bacterial structures in the search to better the performance of presently used natural bacteria.

An excellent overview of the present status is given by Brierley (55), who delves into the as yet relatively little known abilities of bacteria in extractive metallurgy of ores. Of particular usefulness are a number of microorganisms which in effect perform leaching operations with metallic ores. The author envisions a vastly expanded use of the techniques.

ADDENDUM

As previously stated, the abundance of activities is rapidly increasing and so are the writings of investigators. A very good example is that of the Februay 11, 1983 issue of *Science*. This journal is devoted entirely to biotechnology, with 22 articles covering the gamut of subjects. Of particular interest is a report by Johnson (p. 632) on the status of Eli Lilly's commercial production of human insulin, marketed as "Humulin."

REFERENCES

1. *Genetic Engineering News*. Mary Ann Liebert, Inc., Publications. A trade magazine established Jan./Feb., 1981 with Volume 1, Number 1.
2. Sittig, M, Noyes, R. *Genetic Engineering and Biotechnology Firms, U.S.A.-1981*. Sittig and Noyes, 84 Main St., P.O. Box 75, Kingston, NJ 08528.

2a. *Chemtech 12*(9), 516 (1982); under Heart Cut."

3. Anonymous, *Chem. Week* 128 (9): 41–42 (1981).
4. Itakura, K et al. Expression in *E. coli* of a chemically synthesized gene for the hormone somatostatin. *Science* 198 (4321):1056–1063 (1977).

5. Old, RW, Primrose, SB. (1980). *Principles of Gene Manipulation. Studies in Microbiology*, Vol. 2, University of California Press, Berkeley.
6. Losse, G, Mätzler, G. Functional substitutions in the insulin molecule. *Zeitschrift für Chemie* 21(11): 394–403 (1981).
7. Anonymous news item. *Chem. Eng. News* 60 (15):9 (1982); also *Chem. Eng.* 89(8):19 (1982).
8. Dolan-Heidinger, J. Bulletin on Recombinant DNA and biosynthetic human insulin. Eli Lilly and Co., March 1981.
9. R. Crea et al. Chemical synthesis of genes for human insulin. *Proc. Natl. Acad. Sci. (USA)* 75(12): 5765–69 (1978).
10. Chance, RE et al. (1981). The production of human insulin using recombinant DNA technology and a new chain combination procedure. *Peptides: Synthesis–Structure-Function*. Pierce Chem. Co. pp. 721–728.

10a. Aharonowitz, Y. Cohen G. The microbiological production of pharmaceuticals. *Sci. Am.* 245 (3): 141–152 (1981).

11. Goeddel, DV et al. Direct expression in *Escherichia coli* of a DNA sequence coding for human growth hormone. *Nature* 281(5732):544–548 (1979).
12. Olson, KC et al. Purified human growth hormone from *E. coli* is biologically active. *Nature* 293(5831):408–11 (1981).
13. Waldrop, M. Interferon production off to a good start. *Chem Eng. News* 57 (32):24–28 (1979).
14. Office of Technology Assessment (1980). *Genetic Technology: A New Frontier*. Westview Press, Boulder, CO.
15. Nagata S et al. Synthesis in *E. coli* of a polypeptide with human leukocyte interferon activity. *Nature* 284(5754):316–320 (1980).
16. Guron JB. Injected nuclei in frog oocytes: fate, enlargement and chromatin dispersal. *J. Embryol. Exp. Morphol* 36(3):523–540 (1976).
17. Kressmann A et al. Transcription of cloned tRNA gene fragments and subfragments injected into the oocyte nucleus of *Xenopus laevis*. *Proc. Natl. Acad. Sci. (USA)* 75(3):1176–1180 (1978).
18. Edge, MD et al. Total synthesis of a human leukocyte interferon gene. *Nature* 292 (5825): 756–762 (1981).
19. Mantei N et al. The nucleotide sequence of a cloned human leukocyte interferon cDNA. *Gene* 10(1):1–10 (1980).
20. Twigg AJ, Sherratt D. Trans-complementable copy-number mutants of plasmid Col El. *Nature* 283(5743): 216–218 (1980).
21. Bodde T. Interferon: will it live up to its promise? *BioScience* 32 (1):13–15 (1982).
22. Friedman RM. (1981). *Interferons, a Primer.* Academic Press New York.
23. Weissmann Ch. The Cloning of interferon and other mistakes. *Interferon 1981*, Vol. 3, Gresser, I (Ed), pp 101–133.

23a. Hoffman S. Interferon induction (in German). *Z. Chemie* 22(10):357–367 (1982).

24. Fulton SR. Use of Amicon products in production and purification of interferons. *Amicon Application Notes* Publication 552, Amicon Corp., Danvers, MA (1982).

25. Gray PW et al. Expression of human immune interferon cDNA in *E. coli* and monkey cells. *Nature* 295(5847): 503–508 (1982).
26. Gray PW, Goeddel DV. Structure of the human immune interferon gene. *Nature* 298(5877):859–863 (1982).
27. Reyes, GR et al. Expression of human β-interferon cDNA under the control of a thymidine kinase promoter from *Herpes simplex* virus. *Nature* 297 (5867): 598–601 (1982).
28. Hopwood, DA. The genetic programming of industrial microorganisms. *Sci. Am.* 245(3): 90–102 (1981).
29. Anonymous news item. *New Scientist* 92 (1284): 785 (1981).
30. Choo, KH et al. Molecular cloning of the gene for human anti-haemophilic factor IX. *Nature* 299 (5879): 178–180 (1982).
31. Sherwood R, Atkinson T. Genetic manipulation in biotechnology. *Chem. Ind.* (7):241–247 (1981).
32. Zukerman AJ. Developing synthetic vaccines. *Nature* 294 (5845): 98–99 (1982).
33. Dreesman GR et al. Antibody to hepatitis B surface antigen after a single inoculation of uncoupled synthetic HBsAg peptides. *Nature* 295(5845): 158–160 (1982).
34. Edman, JC et al. Synthesis of hepatitis B surface and core antigens in *E. coli. Nature* 291 (5815):503–506 (1981).
35. Valenzuela P et al. Synthesis and assembly of hepatitis B virus surface antigen particles in yeast. *Nature* 298(5872):347–350 (1982).
36. Bittle, JL et al. Protection against foot-and-mouth disease by immunization with a chemically synthesized peptide predicted from viral nucleotide sequence. *Nature* 298(5869):30–33 (1982).
37. Boothroyd JC et al. Molecular cloning of foot-and-mouth disease virus genome and nucleotide sequences in the structural protein genes. *Nature* 290 (5809):800–802 (1981).
38. Küpper H et al. Cloning of cDNA of major antigen of foot-and-mouth disease virus and expression in *E. coli. Nature* 289(5798):555–559 (1981).
39. Kimber H, Stahl S, Küpper H. Production in *B. subtilis* of hepatitis B core antigen and of major antigen of foot-and-mouth disease virus. *Nature* 293 (5832):481–483 (1981).
40. Roitt IM. (1974). *Essential Immunology*, 2nd ed., Blackwell Scientific, Oxford, p. 54.
41. Leder Ph. The genetics of antibody diversity. *Sci. Am.* 246(5):102–115 (1982).
41a. Eisen HN. (1974). *Immunology*. Harper & Row, Hagerstown, MD.
42. Phaff HJ. Industrial microorganism. *Sci. Am.* 245 (3):76–89 (1981).
43. Howard JC et al. Monoclonal antibodies as tools to analyze the sereological and genetic complexities of major transplantation antigens. *Immunol. Rev.* 47:139–174 (1979); also Milstein C et al. Monoclonal antibodies and cell surface antigens. *Cell Biol. Intern. Rep.* 3:1–161 (1979).
44. *Journal of Immunological Methods*, 39 (4): Special Issue (1980). "Special Hybridoma Issue," Elsevier/North Holland, Amsterdam.

45. Rawls R. Monoclonal antibodies key investigative tool. *Chem. Eng. News* 59(50):28–31 (1981).
46. Barnstable, CJ. Monoclonal antibodies which recognize different cell types in the rat retina. *Nature* 286(5770):231–235 (1980).
47. Lee HU, Kaufman SJ. Use of monoclonal anti-bodies in the analysis of myoblast development. *Dev. Biol.* 81:81–95 (1981).
48. New England Nuclear Bulletin. Monoclonal Antibody Screening. Sept. 1982.
49. Hellström KE, Brown JP, Hellström J (1980). Monoclonal antibodies to tumor antigens. In *Contemporary Topics in Immunobiology*, Vol. 11: Warner, NL (Ed.), Plenum Press, New York, pp. 117–135.
50. Bernstein ID, Tam MR, Nowinski RC. Mouse leukemia: Therapy with monoclonal antibodies against a thymus differentiation antigen. *Science* 207(4426):68–71 (1980).
51. Blythman HE et al. *Nature* 290 (5801):145–146 (1981).
52. Peng WW et al. Development of a monoclonal antibody against a tumor associated antigen. *Science* 215(4536):1102–1104 (1982).
53. Lerner TJ. Monoclonals reverse kidney rejection. *Genetic Eng News* 1(5): 3 (1981).
54. Angier, N. The organic computer. *Discovery* 3(5):76–79 (1982).
54a. Robinson AL. Nanocomputers from organic molecules? *Science* 220 (4600):940–942 (1983).
55. Brierley CL. Microbiological mining. *Sci. Am.* 247(2):44–53 (1982).

11
Technology and Design

TECHNOLOGY ASPECTS

There are two technological considerations which will be of chief interest to the chemical technologist and, specifically, the chemical engineer.

One should be aware that some, even if limited, knowledge of gene manipulation is essential. Thus, it will be most helpful to acquire such knowledge through instruction in biochemistry or microbiology, preferably in both areas, or in biochemical engineering which deals mainly with fermentation processes.

The two areas of greatest concern to designers are equipment design and process development. Equipment engineering would appear to already have been solved for the most part in existing biochemical industries. Whereas the present state of gene manipulation procedures is mainly on the laboratory scale, there are situations where innovative concepts need to be incorporated into designs of some new equipment needs. Initially, the scaling-up may have to be done by "brute force" while adequate designs are being developed.

Equipment difficulties as well as operating efficiency will be improved by effective process engineering. Some of the laboratory procedures, for instance, plasmid DNA isolation (or viral DNA) are not at all suitable for scaling-up to plant operation. It is this area wherein the technologists will find their greatest opportunity for devising process improvements.

The development of penicillin manufacturing is a very instructive example of what can be accomplished through research and development. Penicillin mold originally was obtained from its growth as a green slime on the surface of a nutrient solution–naturally in farm ponds and commercially in laboratory cultures. After Fleming's discovery of this "antibiotic," efforts to produce it commercially moved rather slowly. World War II suddenly sent demands sky rocketing, and penicillin production was scaled up by "brute force."

Woodframe scaffoldings were installed in warehouse facilities to hold 5-gallon carboys. These were charged with nutrients, laid sideways, and topped off to the bung hole. Culture was then implanted and the mold observed in surface growth. When spot assays indicated that penicillin production was at the desired level the floating mycelium was "fished" out through the bottle neck. From this point on the extraction of penicillin followed conventional lines. Finally, the relatively pure penicillin was finished by freeze drying (lyophilization).

The laborious surface growth method obviously was unwieldy and vigorous efforts in process development resulted in the isolation of a mold strain which would grow and reproduce in *submerged fermentation*. This was the beginning of economical mass production of penicillin and similar antibiotics. The cost of antibiotics to the patient was reduced at least a thousandfold.

TECHNOLOGY: LABORATORY SCALE

The descriptive section of U.S. patent 4,237,224 (1) entitled "Process for Producing Biologically Functional Molecular Chimeras" contains a wealth of information. The inventors are S.N. Cohen and H.W. Boyer. The patent, a process patent, is assigned to Stanford University, which established a rather comprehensive licensing policy. The "Stanford University Patent License" was published in its entirety in *Genetic Engineering News* (2). As one would expect, the patent language is not casual reading and at least some degree of familiarity with the subject matter is needed to benefit from the information.

In any case, steps in undertaking a recombinant DNA procedure require:

1. Sources of vectors such as plasmids or viral DNA (phages)
2. Restriction enzymes for specific site cleaving of DNA
3. Sources of DNA segments which contain a desired gene sequence
4. Ligating enzymes to establish plasmid DNA closure to circular state
5. Hosts to receive the plasmid or phage chimera so that they can replicate and produce the desired component (proteins, insulin, interferon, etc.)

A "numerical" example is described in considerable detail in the *above cited patent* (1). The following is a schematic outline of the procedure to illustrate the "processing" sequence. For the sake of brevity the detailed quantitative values are not cited.

a. *Preparation* of pSC101 plasmid (code number, calculated MW = 5.8×10^6); covalently closed R6-5 DNA was sheared in a microblender at 2000 rpm for 30 min in a TEN buffer solution at 0° to –4°C. (*Note*. this is a mechanical procedure for shredding DNA–an appropriate restriction enzyme could have been used).

b. The sheared DNA was separated into fractions by sucrose gradient sedimentation at 39,500 × g and 20°C. A fraction (it does not specify where it was isolated in the gradient) was collected on filter paper, dried for 20 minutes, and precipitated by immersing the filter disc in cold 5% trichloroacetic acid (TCA) containing thymidine. The precipitate was filtered, washed once with TCA, and twice with 99% EtOH, and dried.

c. The quantity of chimera DNA needed for insertion into bacteria and cloning is very small, perhaps 0.5-5 μg. After insertion, the bacterial growth will produce adequate amounts to allow culturing large batches. Therefore, processwise, one will have an appropriate microbiological service laboratory as the source of cultures for plant-scale production. This means that presently available equipment can be used. Rather than mechanical shearing of the DNA, the technologist might prefer enzyme cleaving, with an appropriate endonuclease. Such a cleaving procedure permits creating so-called cohesive or "sticky ends" which facilitate subsequent ligation to the circular configuration. The real problem is to devise schemes to transform the minute laboratory scale into a commercial operation.

DESIGN CONSIDERATIONS

To make recommendations on the design of genetic engineering plants, pilot-scale, or commercial unit, without some actual experience, is somewhat presumptuous. Very few operations so far have gone past the biochemistry laboratory stage and thus pertinent "experience" is still a rarity.

Nevertheless, there are related areas such as the fermentation industries (biochemical engineering), the pharmaceutical operations, and food manufacturing, and these areas are concerned with many of the problems which will be encountered in genetic engineering plants. As a matter of fact, many of the truly genetic engineering procedures may be done on a relatively small laboratory scale, to wit the extraction of plasmids from the bacterial host, the preparation of the plasmid by gene insertions (see e.g., Chap. 7, Fig. 2), and the introduction of the "manufactured plasmid" into a bacterial host for cloning. At this stage, scaling-up can begin as the bacterial host organisms will grow and multiply as in any other fermentation step. It should be realized that 1 μg of recombinant DNA can yield as many as 10^5 different transformants or clones. Thus, minute quantities of DNA can be sufficient to obtain the desired bacterial clone, and scale up will then occur to produce the desired product from that clone.

The January 1982 issue of the *Journal of Chemical Technology and Biotechnology* (3) contains a wealth of pertinent information on problems arising in biochemical operation. Of particular interest is a section on Bioreactor Design and Performance.

Logistics of Scaling-up Problem

The chemical technologist takes at least gram quantities for granted in laboratory work, and kilograms, pounds, gallons, barrels, or tons in manufacturing. The recombinant scope of quantities, well below the gram scale, then presents a new dimension in quantitative thinking. Thus, it will be helpful as an exercise in logistics to consider a flow sheet which represents the "purification of milligram amounts of DNA fragments" (4). Although it is not necessary to obtain milligram quantities of DNA to obtain a particular clone producing a desired product, the production of large quantities of popular cloning vectors can be a viable commercial objective. Companies such as BRL (Bethesda Research Laboratories), for example, sell pBR322, pUC8 and pUC9, and λ phage vectors.

To begin, the milligram level already represents a 1000-fold scale-up of the more usual microgram level. While this procedure, purification of milligram quantities, may not have to be scaled up to semiplant or commercial size, it is likely that multigram or even kilogram amounts will be needed for recombinant work leading to the cloning steps. So, a trip through the biochemical labyrinth of the process will be enlightening. It is most easily envisioned by means of a process flowsheet.

The diagram was prepared from a set of fairly detailed directions published by Hillen, Klein, and Wells (4). Their work involved "Procedures for rapid preparation of gram quantities of pure recombinant plasmid DNAs, selective polyethylene glycol precipitation of DNAs in different lengths, and large scale high-pressure liquid chromatography. . . ."

The "biochemistry" flow diagram, as developed by a group assignment in design, contained the surprising number of 90 procedural steps (not counting several multiple batch extractions). Subsequent conversion to a "technology" diagram caused considerable discussions, particularly whether batch or continuous operation should be selected. The final "compromise" opinion opted for batch procedures with all-glass equipment. There are several suppliers of process glass units, designed on a unit operations principle. A worthwhile reference is a catalog by the Owens-Illinois company, which supplies modular systems developed by the German "Schott" corporation.

The main purpose of the presentation of the minutia involved in the biochemical laboratory procedure is to impress the technologist with the scale of operation and multiplicity of operations which must be converted into a workable plant scheme. For instance, what quantities of DNAs will be needed to effectively conduct recombinant work for insertion in vectors and subsequent cloning? Will it be more economical to provide a larger number of fermentation tanks and keep the recombinant operations at a laboratory scale? Also, a flowsheet should be of considerable help in translating the biochemist's report into a meaningful picture of the task which is faced in design.

A systematic approach would entail the following steps:

1. Analysis and discussion of the report prepared by biochemistry experts.
2. Preparation of detailed flowsheet incorporating all biochemical procedures.
3. R&D analysis of the flowsheet for possible condensation of steps and simplification of sequential operations.
4. Economic decisions on which steps can be kept at a laboratory scale, perhaps scaled-up to bench scale, and what operations must be developed to full-scale status.

DESIGN CLASSIFICATION

There are two design categories which call for different approaches in principle, but they also are interrelated and have to be fused in the overall design. Thus, process design will determine the various operations which have to occur sequentially and equipment design must blend the hardware into the manufacturing scheme.

Process Design Items

Scale of Operations

As presently conducted the scale of operation in the laboratory procedures is extremely small. A news-type article in *Chemical Week* (5) covers some interviews with knowledgeable industry experts. It is a rather brief account, but contains a great deal of helpful information on the major problems which are likely to arise in the new DNA field. A production of 1×10^{-14} g per experiment is quoted, but this figure appears to be much too low. Actually, microgram amounts are more likely. This would agree with another estimate which states that genetic key insertions are accomplished in the 1 mg range, while bench top fermenters can produce 100 to 1,000 g/day. Compare this to what would be a small output for a chemical pilot plant of perhaps 100 kg/day, and at best it would still be only 1/100 of that production rate.

However, commercial production is reckoned on a tons/day basis and so the scale-up problems are highly magnified. Commonly, 10,000 gallon fermenters are somewhat of a standard size for pharmaceutical manufacture, and 100,000 to 150,000 gallon capacity are stated to be in use in penicillin manufacture. Obviously, as far as fermentation type of procedures are concerned, which in genetic engineering would be the cloning step to grow the host organisms, it is safe to say that the technique exists for commercial-scale operation. The largest so-called *biotechnology* installation is said to be the Imperial Chemical Industries 130 million g/day unit which produces single-cell protein.

Processing Steps

Fluid Flow

Problems in this area will not be unique. The pumping of microbial suspensions is well in hand in fermentation processes. When DNA and RNA become involved, then essentially it is a matter of handling solutions, provided the pH value is kept at the proper level. Some further comments will be made under the heading of "piping."

One can expect that data on flow properties of microbiological suspensions are available as in-house information. Apparently, however, published information is somewhat scarce. So, it is of interest that a recent publication by Reuss et al. (6) presents experimental results on the rheological properties of fermentation fluids. An important aspect is that flows in such systems are affected by the osmotic pressure of the biological entities.

Heat Transfer

Generally, operating temperatures will be in the 20° to 40°C range with occasional heat-shock treatments and also some temperature cycles to perhaps -20°C. However, these cycles may possibly occur only in laboratory operations, when recombinant gene insertions are carried out. Processing at the more normal temperatures will occur in fermentation, that is, cloning stages. Overall, however, attention must be paid to the fact that the processing involves biological entities which are quite heat sensitive.

Separation Operations

In laboratory practice centrifugation is often used, as well as filtration to remove plasmids, DNA and RNA from suspensions. Note that DNA and RNA can readily be precipitated from their solution by addition of NaCl and alcohol. Again, most of these steps are more likely in the laboratory preparation stage. Separation of cells in suspension is unlikely to be achieved by settling, as the cells are too buoyant, and so continuous centrifugation or moving-belt filtration has to be used with large-scale cloning.

Extraction

When the bacterial clones have produced the desired compound, for instance an insulin protein chain, the cells will have to compacted and then ruptured, either mechanically or by lysing with enzymes. Extraction of the product then will generally call for filtration to remove cell debris, and the use of a specific extraction procedure such as liquid-liquid, membrane diffusion, etc. Also, for the more valuable materials, such as enzyme proteins and nucleic acids, the use of aqueous gel filtration chromatography might be considered (7).

Sanitation

Sterilization practice would be as customary in any of the pharmaceutical, food, and fermentation plants. While DNA and RNA as such do not require sterile handling because they are just chemicals, one would strive for sterility to avoid contamination and to remove nucleases which would degrade the nucleic acid. Also, provision must be made for sanitary disposal of microbial waste, which should include processing to kill live bacteria.

It is important to realize that design objectives for assuring proper functioning of process sterility must be an innate part of initial planning. A statement by Challenger (8) is worth quoting:

> Clear lessons learnt to date include the fact that it is virtually impossible to impose sterility of feed solutions, air supply and equipment on a *conventional* plant design. These requirements must be integrated with the basic process design from day one, the particular process determining the standard of sterility required. This is not just a question of avoiding direct connections between sterile and non-sterile areas as bacteria have been known to grow through conventional valves. Specially designed valves may well be needed, each having to justify its existence.

Challenger's knowledgeable discussion covers a multitude of essential equipment and process precautions and should be consulted directly.

Obviously, there will be problems in disposing of microbial wastes, occasional clean-out and sterilization of lines and equipment, clean-up of laboratory wastes and correction of spot contaminations in equipment. Certainly, steam sterilization will be used wherever it is convenient and available. Other incidents may call for liquid sterilants such as 0.02% sodium azide (*no contact* with lead, copper, or other azide formers!), 0.005% trichlorbutanol in weakly acidic solution, 0.005% merthiolate (weakly acidic), 0.002% chlorhexidine and 0.001 to 0.01% phenyl mercuric salts (weakly alkaline) (9). Gas sterilization with ethylene oxide may at times be possible; however, it is very soluble and may be retained long after sterilization is completed. A particularly undesirable feature is its retention in plastic compositions.

A common procedure in recombinant techniques is the solvent extraction of protein from DNA solutions by means of phenol. Therefore, there is the added problem of removing proteins from the phenol solution to recycle the phenol. Obviously, the proteins will be thoroughly denatured and will not present disposal problems, provided residual phenol content will be kept to a very low level.

EQUIPMENT DESIGN FEATURES

The nature of the processed materials is that of living microbial entities. Some at least are rather fragile organisms and most are quite heat sensitive. Such properties impose serious restrictions on the handling of plant streams and reactor contents so that stream transportation and reactor operation call for special precautions.

Reactors and Heat Exchangers

One should expect that microorganisms will behave differently in large vessels than in the test tube. How differently, only experience will tell. There are at least two major differences from standard chemical manufacture. First, almost all biological reactions are exothermic. Second, the temperature level of the reactions, such as cloning, is essentially ambient, say 25° to 37°C. This means that the heat liberated, which will also vary with the microbe population, must be removed at a low temperature level and most likely with a rather small temperature differential. Also, the low temperatures mean that the reaction rates are slower and time cycles present an additional control problem.

In the average chemical plant there will be a single heat exchanger large enough to handle design loads for a particular unit, and the temperature differential will be high enough to provide flexibility. In a microbial reactor, with engineered plasmids within hosts, it probably would be well to use several smaller heat exchangers, so that a large but variable heat transfer area is available to match up with the changing heat load and the low Δt's, keeping in mind that the reactor heat generated is proportional to the reactor volume, while the heat removal is controlled by surface area. The problem of *adequate agitation* within the reactor is of prime importance, as it affects reaction rates and heat generation (10). A conventional batch reactor set-up is shown in Figure 1. As customary, the mixing is attained by means of a series of paddles (impellers) on a rotor shaft. A recent report on a bubble-column with outside recirculation of reactor broth contains novel data for an *E. coli* culture (11) Such equipment may provide some attractive design features. A novel agitator design was described by Sittig and Faust (12). It is called an "airlift loop reactor" and its principle is illustrated in Figure 2. Test data show that 95% homogeneity in the broth can be attained in only 5 passes through the updraft section. This was accomplished in about 100 seconds in a 27 cu m^3 volume reactor. A pertinent discussion of

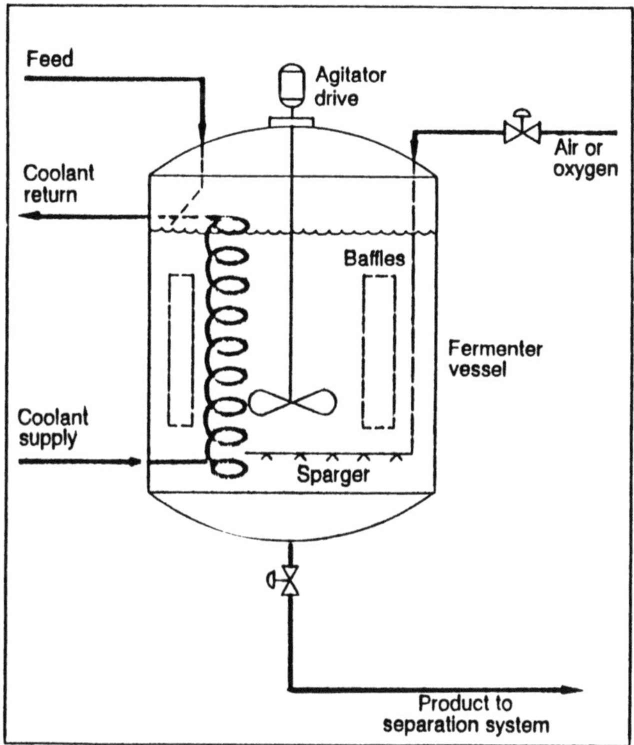

FIGURE 1 Conventional batch reactor. Most frequently used type of reactor in microbiological plants.

design parameters in microbiological reactors was presented by Andrew (13), who also covered the problems encountered in scale-up from laboratory data.

Piping, Valves, and Pumps

With a reactive milieu containing living microbes, it is important to pay special attention to the possible occurrence of dead spaces. Thus, deadend structures, sharp bends, pipeline dips, and in-line valve projections should be avoided. Valves should be of the throughflow type; butterfly valves may be satisfactory, but the possibility of culture adhesion and growth on the valve seat must be considered. A highly desirable feature would be three-way valving arranged to permit line blowout when circulation is shut down. Microbes in dead-space will grow and produce complete line blockage.

Pump type should be selected for smoothness of the flow profile, thus

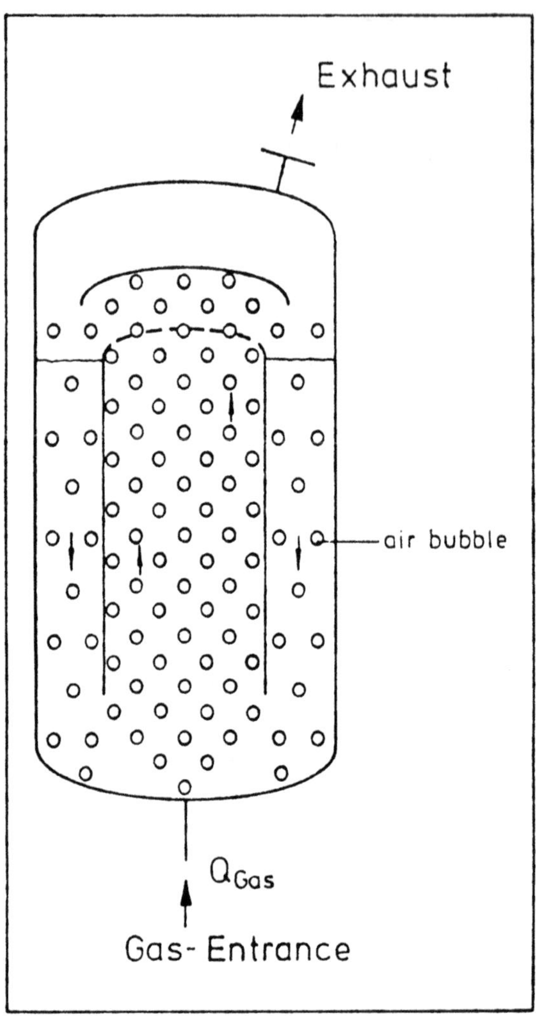

FIGURE 2 Airlift loop reactor. Principle of a gas-agitated reactor with internal recirculation of broth. (Courtesy *The Chemical Engineer*, Ref. 12).

centrifugal and membrane pumps might be preferred. Roller pumps working by plastic tube compression (squegee pumps) could rupture an unacceptable number of cells.

Materials of Construction

Customarily, the materials used in related industries are stainless steel, aluminum, glass, and ceramics. While polymer equipment and polymer lining may look attractive, there is the possibility of compound leaching of occluded monomer, placticizers, and stabilizers. Such constituents could prove toxic to the cultures. Furthermore, many microbes may attack plastic compositions. One must also consider that some of the biochemical solutions may have a fairly high salt content, and usually will be well oxygenated so that stainless steel could be subject to corrosion.

Instrumentation

For the most part, present instrumentation techniques should take care of the control problems in genetic engineering plants. Of course, there will arise some unusual situations and appropriate design modifications of existing hardware may be necessary. An unknown factor as yet is the amount of instrumentation that will be needed and consequently a high cost factor can ensue. Thus, the previously quoted survey (5) contains the statement that ICIs single-cell protein plant requires 1,000 pH meters which have to function in a rather aggressive culture broth.

Siebert and Hustede (14) report a study on submerged fermentation for the production of citric acid in plant size fermenters to evaluate ultimate computerized control of operation. A plant fermenter was equipped with extensive instrumentation to permit on-line collection of operating parameters. It was established that microbial process control could be improved substantially. Also, the considerable wastewater burden associated with this type of fermentation could be brought to a more manageable level. The success of this computer-controlled operation could find profitable use in the cloning phases of genetic engineering. Information on computer hardware and software is given in sufficient detail so that a direct carryover should be possible.

Potential Recovery Scheme

A pertinent analysis of downstream processing in the biochemical technology was presented by Hawtin (15) in the British journal, *The Chemical Engineer*. Hawtin's presentation covers the overall problem, that is, cell separation, cell rupture, and subsequent separations and purification steps. A flow diagram is worthy of reproduction (Fig. 3); it is in effect self-explanatory.

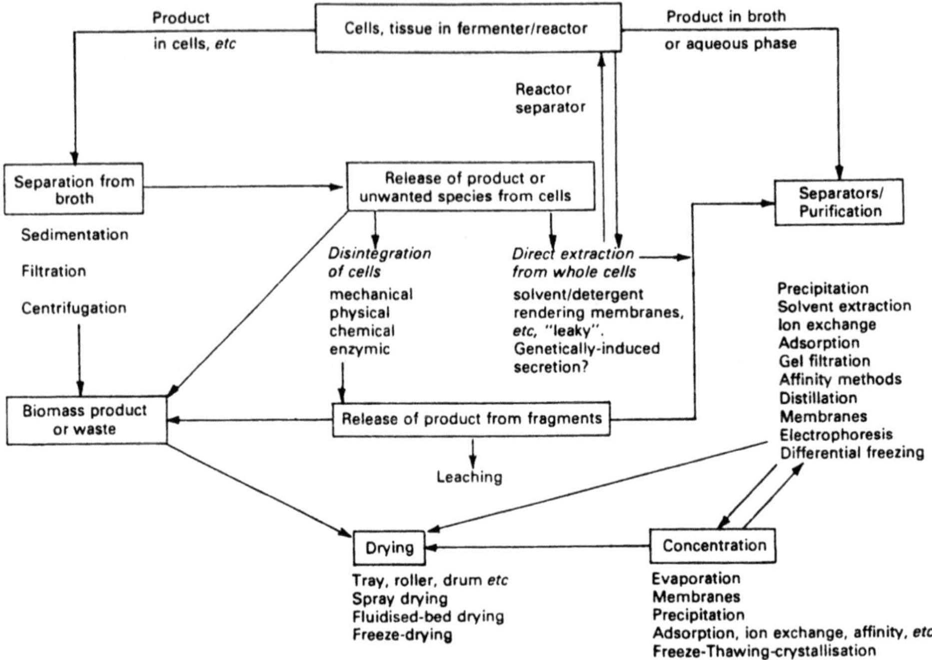

FIGURE 3 Downstream processing scheme. The processing of the product from a biomass reaction by conventional steps. (Courtesy *The Chemical Engineer*, Ref. 15).

As far as is known at present, the generated drug proteins generally remain inside of the cell. To harvest the desired protein products, the bacterial cells have to be broken. Extractive procedures are then used to isolate and purifiy the product as shown in Figure 4. A very significant development was given as a news item in *Chemical Week* (March 31, 1982). It states that workers at Genentech have spliced *human interferon* gene into yeast cells and that the *cells secrete* at least some of the interferon (10–20%) into the surrounding growth medium. One can envision that the excreted interferon can be continually removed by circulating the broth to an external extraction unit and returning the interferon-depleted stream to the fermentor. Such a procedure should result in a continuing concentration gradient between the cells and the broth, leading to further diffusion of manufactured proteins from the cell to its surroundings.

The Membrane Reactor

The utilization of a selective permeable membrane in conjunction with a chemical reaction was treated by Hwang and Kammermeyer (16) under the heading "coupling with chemical reactions." More recently the term "membrane reactor" has come into use. The basic concept is the selection of a suitable membrane as one of the confining walls of a reactor. Specifically, such a membrane must be preferentially permeable to the desired reaction product which then is continuously removed from the reaction mixture. When this is achieved, the reaction will continue in the direction of product formation, resulting in excellent conversion and at the same time it presents a separation and purification process directly coupled to the reactor.

An important and actually critical element is the selection of the membrane, both materialwise and in regard to structural characteristics. A valuable, up-to-date treatise by Lonsdale (17) will be most helpful in this matter.

An obvious prerequisite for the use of such a scheme is the presence of the reaction product so to speak in a free state to have access to the membrane. In the usual case of biological cloning and associated product formation this is not the case. Actually, the cell wall of the host bacterium would represent the permeable membrane and the cell itself, the reactor. Only, if the cell wall were to allow outdiffusion of the desired product (insulin protein chains, HGH, somatostatin, interferon, etc.) would there be a readymade "natural membrane reactor."

Membrane-Bound Cells

In recent years, the technique of immobilizing bacterial cells on or in a microporous medium, has received continual attention. The technique is in commercial use and R&D activity is intense. Michaels (18) presented a comprehensive overview with particular attention to fermentation practice, the process which will be the cloning step in recombinant applications. The membrane reactor principle, as well as immobilized culture growth are elucidated. It is evident that such techniques will be of use only if the cell will diffuse its product into the surrounding liquid medium. In such a situation, recirculation of liquid containing nutrients on the one hand and leaching production on the other hand, would constitute an elegant continuous process.

The possibility that bacterial cells will allow diffusion of such reaction products through their cell wall into the surrounding liquid phase medium was mentioned in the preceding section. This situation is then ideally suited for further downstream processing. A very important investigative aspect would be to screen for bacterial hosts which are capable to act as a "native membrane reactor"—undoubtedly a formidable task. There is one other approach which deserves attention. It is the use of a diffusional promoter in the reaction medium (the broth) which would enhance product diffusion through the cell wall. Obviously, there are some stringent limitations. Diffusional promotion could for instance be attained by the presence of an organic solvent as a "carrier" of the product. However, the effect of such a solvent on the viable organism could well be disastrous. Yet, there surely are some suitable promoters which will be tolerated by the cellular entity. True, not an easy task, but one worth pursuing.

Fixed-Bed Processing

What may become an important procedure in cloning is a process that utilizes microporous fibers as in situ media for bacterial growth. Inloes et al. described an in-depth study of the solid phase-bacterial system (19). Experimentally, *E. coli* bacteria were grown in the macropores of hollow fiber matrices to a population density of some 10^{12} cells/ml of accessible void volume. The performance of the biosynthetic system was traced by the production rate of beta-lactamase (a protein) which served as an indicator of cell productivity.

The plasmid was pBR322 in *E. coli* strain C600. The hollow fiber membranes were of the asymmetric wall type as used in ultrafiltration. They provided the macroporous layer accessible to the bacterial culture. The design of the reactor is sketched in Figure 4. The growth pattern of the cells is pictured in Figure 5 as an electron microphotograph at 2000X. It represents a cross-section of the membrane where the elongated sausagelike areas appear as dark-colored cell masses. The effectiveness of the system was established by comparison with conventional shaker-flask culturing. Overall, the per cell productivity of the reactor was considerably less than in shaker flasks, but on a reactor-volume basis the reactor was 100 times more productive than the shaker-flask cultures.

There is, of course, the inherent limitation that the product produced by the

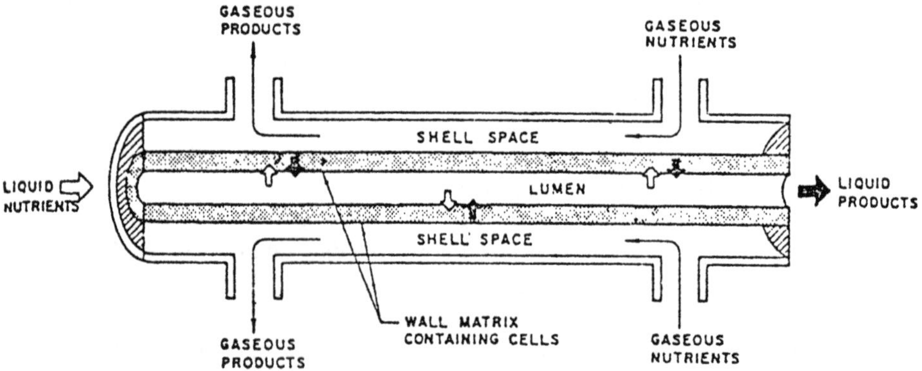

FIGURE 4 Fixed bed reactors. Reactor contains microporous fiber elements in which bacterial cultures are grown. The products produced by the cultures are removed by diffusive leaching with a liquid nutrient solution (Courtesy Biotechnology and Bioengineering)

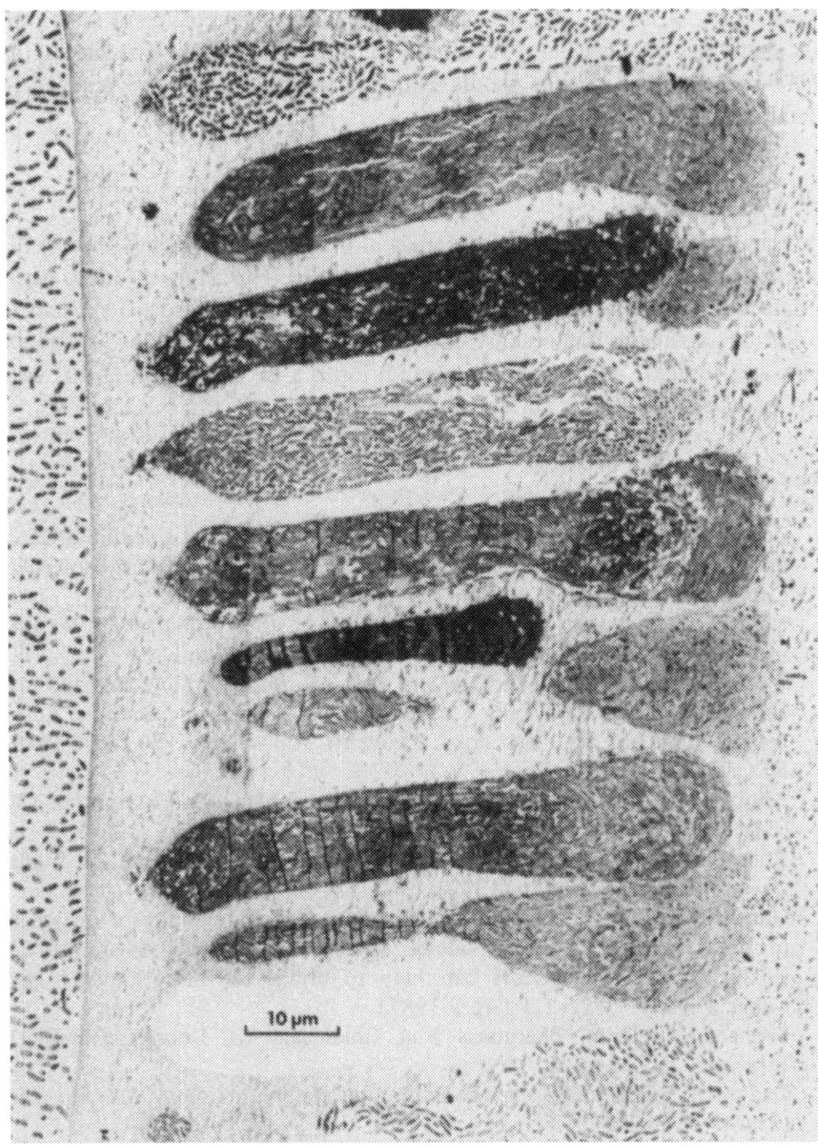

FIGURE 5 Bacterial growth in hollow fibers. Macropores in the matrix of asymmetric-wall hollow-fiber membranes serve as culturing sites for bacteria, specifically *E. coli* (19).

cells must be, at least in part liberated to the liquid-phase portion of the broth. This matter has been belabored repeatedly in previous sections. Nevertheless, the fixed-bed membrane reactor should constitute a highly effective adjunct to recombinant technology.

ADDENDUM

A most valuable reference list of Materials of Construction was published in the October 18, 1982 issue of *Chemical Engineering* (20). The listing covers 20 pages of data on design properties, codes, design calculations, corrosion resistance, fabrication, standard dimensions, cost data, technical description, reference tables and sources. A very convenient addition is a list of companies that supplied information and their current addresses.

REFERENCES

1. Cohen SN, Boyer HW. U.S. Patent 4,237,224, assigned to Leland Stanford, Jr. University, "Process for Producing Biologically Functional Molecular Chimeras," December 2, 1980.
2. *Genetic Engineering News*. Mary Ann Liebert, Inc., Publications. A trade magazine established Jan./Feb. 1981 with Volume 1, Number 1.
3. *Journal of Chemical Technology and Biotechnology 32* (1):1-364 (1964).
4. Hillen WH, Klein RD, Wells RD. Preparation of milligram amounts of 21 deoxyribonucleic acid restriction fragments. *Biochemistry* 20 (13): 3748-56 (1981).
5. Anonymous. Tiny bugs need to make it big in chemicals. *Chem. Week* 129 (24):42-43 (1981).
6. Reuss M, Debus D, Zoll G. Rheological properties of fermentation fluids. *Chem. Eng.* 381: 233-236 (1982).
7. Gurkin M, Patel V. Aqueous gel filtration chromatography of enzymes, proteins, oligosaccharides and nucleic acids. *Am. Lab.* 14 (1):64-73 (1982).
8. Challengeı JG. Contractors can help to bridge the biotechnology gap. *Chem. Eng.* 379: 127-129 (April 1982).
9. Sephacryl, bulletin, Pharmacia Fine Chemicals AB, Uppsala, (Sweden), Picataway, NJ.
10. Gaden EL Jr Production methods in industrial microbiology. *Sci. Am.* 245 (3):181-196 (1981).
11. Adler I. Schügerl K. Reactive kinetic investigations with *E. coli* in a bubble-column with external recirculation (in German). *Chem Ing. Tech.* 53(12):968-9 (1981).
12. Sittig W, Faust U. Biochemical application of the airlift loop reactor. *Chem. Eng.* 381:230-232 (June 1982).
13. Andrew SPS. Gas-liquid mass transfer in microbiological reactors. *Trans. Inst. Chem. Eng.* 60 (1):3-313 (1982).

14. Siebert D, Hustede H. Citric acid fermentation–biotechnology problems and potential computerized process control (in German). *Chem. Ing. Tech.* 54(7):659–669 (1982).
15. Hawtin P. Downstream processing in biochemical technology. *Chem. Eng.* 376 (1982).
16. Hwang ST, Kammermeyer K. (1975). *Membranes in Separations*. John Wiley & Sons, New York.
17. Lonsdale HK. The growth of membrane technology. *J. Membr. Sci.* 10 (2:3):81–181 (1982).
18. Michaels AS. Membrane technology and biotechnology. *Desalination* 35: 329–351 (1980).
19. Inloes DS et al. Hollow fiber membrane bioreactors using immobilized *E. coli* for protein synthesis. *Biotechnol. Bioeng.* (in print).
20. Materials of construction: current literature. *Chem. Eng.* 89 (21):94–118 (1982).

12
Techniques with Mammalian cells

The introduction of a mammalian gene into a recombinant structure and its reintroduction into mammalian cells requires the use of a viral vector. Most such vectors are derived from the simian virus 40 (SV40), which was discussed in Chapter 7 in connection with the construction of an SV40-λ DNA hybrid (Chap. 7, Fig. 11). The search for and construction of new vectors was described by Mulligan and Berg (1). Several recombinant plasmid vectors are shown there, based on pBR322 and SV40 sequences.

GENE INTRODUCTION TO A MAMMALIAN CELL

A rather illuminating scheme was presented by Anderson and Diacumakos (2). The diagram shown in Figure 1 illustrates the creation of separate plasmids which carry the *TK gene* and the *human beta-globulin gene*, respectively. The procedure begins with the introduction of these genes into receptive bacterial plasmids and subsequent microinjection to place the recombinant plasmids into the nucleus of a mammalian cell. The *TK gene* is responsible for the formation of the enyzme "thymidine kinase": it is obtained from the herpes simplex virus. The absence of the TK gene is a genetic defect in mouse cells, leading to cell death. The human beta-globulin gene is of great importance as mutations in that gene cause sickle-cell anemia and the related disease thalassemia: These are two hereditary disorders of red blood cells.

The upper right area of the diagram shows a fragment of viral DNA, which carries the TK gene, which was isolated by cleaving with *Bam*HI and separated by gel electrophoresis. This fraction is then introduced into a cleaved pBR322 in which the tetracycline resistance was eliminated because the *Bam*HI cut

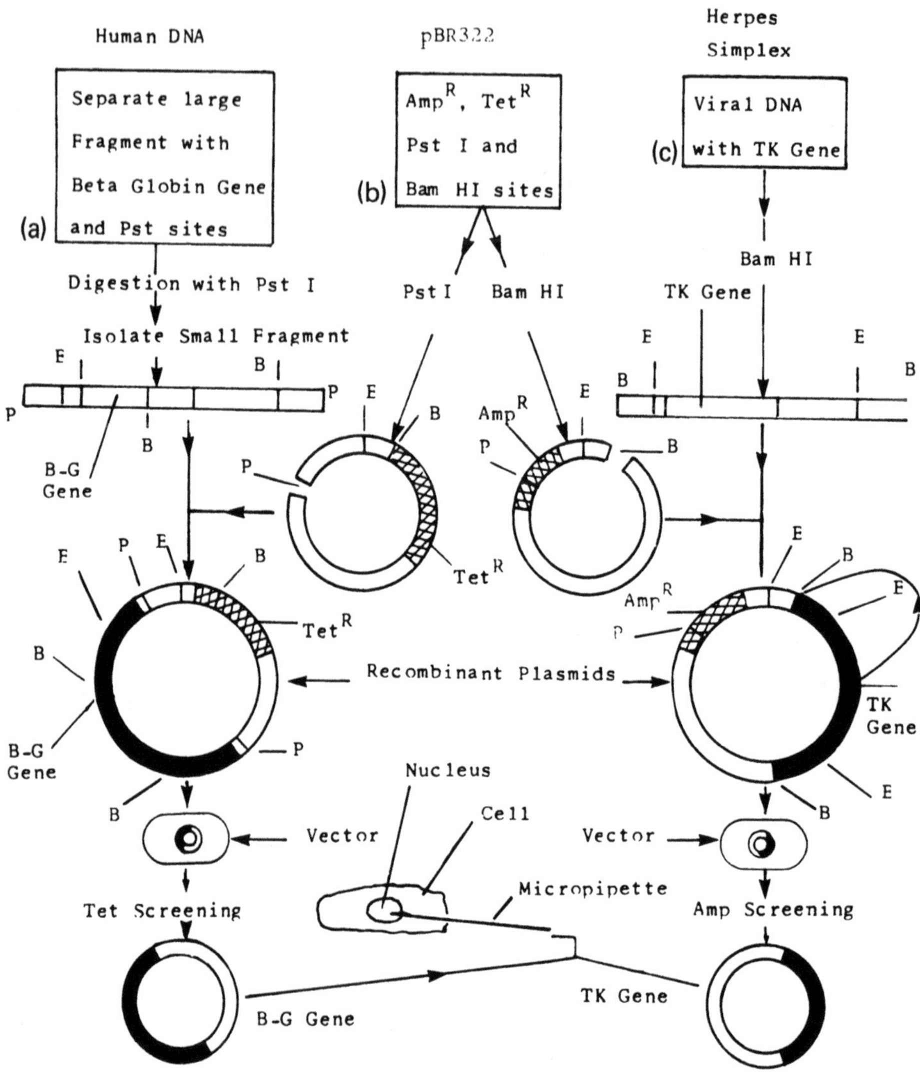

FIGURE 1 Recombinant plasmids with mammalian genes. (a): Separation of beta-globin gene from human DNA. Insertion into a pBR322 plasmid, cloning in vector, and Tet screening to get final recombinant plasmid with BG gene; (c): Isolation of sequence from viral DNA-containing TK gene. Same procedure as in left lane to get recombinant plasmid with TK gene and Amp^R; (b): Cleaving pBR322 plasmids to match up with BG and TK sequences. Final: microinjection into a viable cell. E = *E. coli*; B = *Bam*HI; P = *Pst*I; BG = Beta-globulin gene; TK = Thymidine kinase gene.

through the *tet* gene. Continuing at the right branch, the recombinant plasmid, the result of ligation of viral DNA fragment and cleaved pBR322, is screened on ampicillin to isolate the proper clone. Similarly, in the left branch of the figure, the human DNA is fragmented with *Pst*I, sequenced on gel, and the proper gene fragment is excised from the gel and eluted. Also a pBR322 plasmid is cleaved with *Pst*I to destroy the ampicillin resistance and to linearize it for recombination with a beta-globulin fraction. Ligation again yields a recombinant plasmid, which is then screened on tetracycline and cloned. Finally, the two new structures are microinjected into a mammalian cell to observe if the new genes will be expressed in cell division. The details of the experimental procedures, the technique of microinjection, and an analysis of results are well presented in the original publication.

HUMAN β-GLOBULIN GENE IN A MAMMAL

Attempts to introduce mammalian genes from one source into a different mammal have been beset with many difficulties and often have failed. Thus it is significant that Stewart and co-workers (3) appear to have succeeded to insert human β-globulin genes into mouse eggs. A recombinant plasmid which contained the β-globulin genomic region was injected into fertilized mouse eggs. The plasmid also carried the herpes simplex viral thymidine kinase gene. Both the human and the viral genes were identified in the DNA of one of the animals. Another animal had incorporated only a part of the human globin gene, but it transmitted the gene sequence to its offspring in a Mendalian ratio. The published article presents gel electrophoresis diagrams which substantiate the findings.

HUMAN GENETICS

While the emphasis in the term *genetic engineering* is placed more and more on the development and utilization of *recombinant* techniques, it should not be overlooked that this term was applied many years ago to human genetics. Indeed one of the earliest texts was written by a physician, Dr. Lawrence E. Karp (4), who dealt entirely with the techniques available to evaluate genetic characteristics and their consequences in the human entity. It seems appropriate then to give a brief account of some essential procedures utilized in the determination of human genes. Many readers may welcome such information as it deals with some intimate problems which might be faced by us all.

The "genetic inheritance" process can be conveniently demonstrated by the example of human chromosome characteristics. If, for instance, a tissue biopsy is taken by a swab from inside of the cheek, it can be manipulated with microsurgical techniques under a light microscope to locate cells which are at the proper

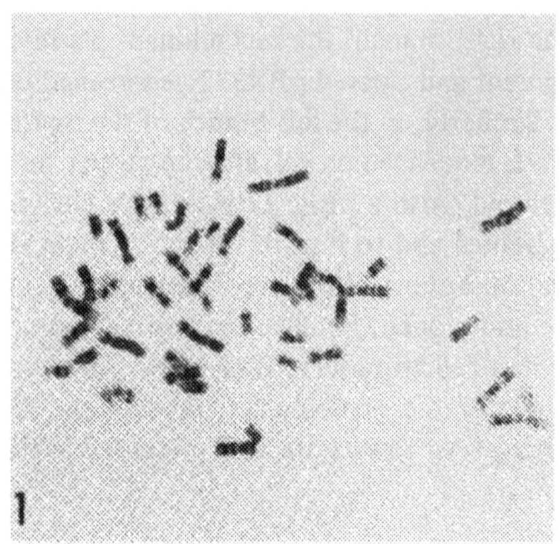

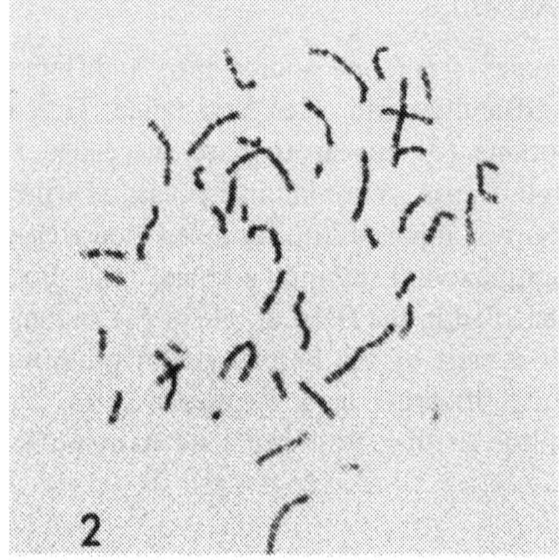

FIGURE 2 Photomicrograph of chromosomes. Chromosomes at metaphase showing various stages of condensation (1–4) before they are cut out for karyotyping. Elongated chromosomes (2–4) are more informative for detecting small abnormalities because they exhibit more bands. (Courtesy Dr. Shivanand R. Patil, Medical Cytogeneticist, College of Medicine, University of Iowa).

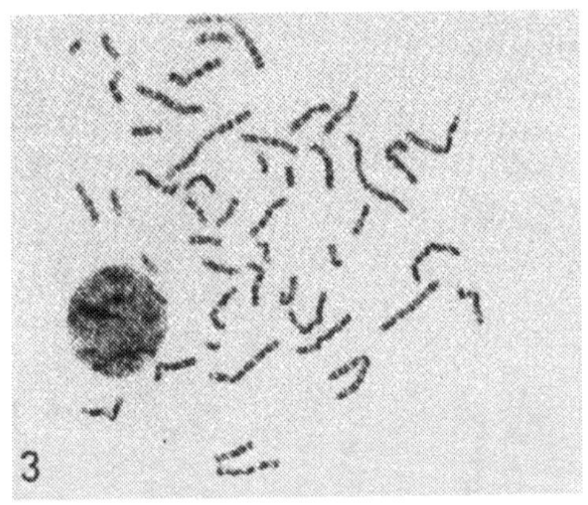

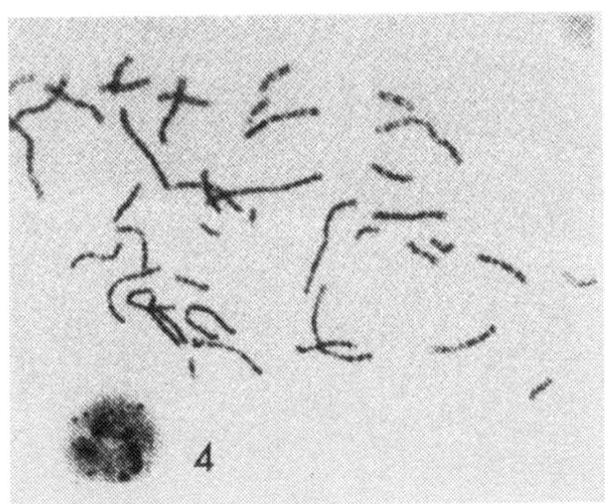

FIGURE 2 (continued)

stage to visualize chromosomes. (*Note*: Because the chromosomes are visible only at cell division, a buccal smear is used to pick up a large number of cells. Several cells are then placed on a glass slide and squashed under the microscope. In this way one or more will be found which are at the proper stage of cell division.)

When the cells burst, chromosomes are released and spread out on the slide. The microscopic view is photographed as shown in Figure 2. After enlargement, the individual chromosomes are cut out and paired according to size and in reference to a constricted region, called *centromere*, (also, kinetochore). Obviously, this organizing (karyotyping) requires extraordinary skill. Each cell of the human body contains 23 pairs of identical chromosomes (a total of 46). The completed arrangement, as shown in Figure 3, is called a *karyotype*. The arrangement is that of a female cell where chromosome 23 is the sex chromosome XX.

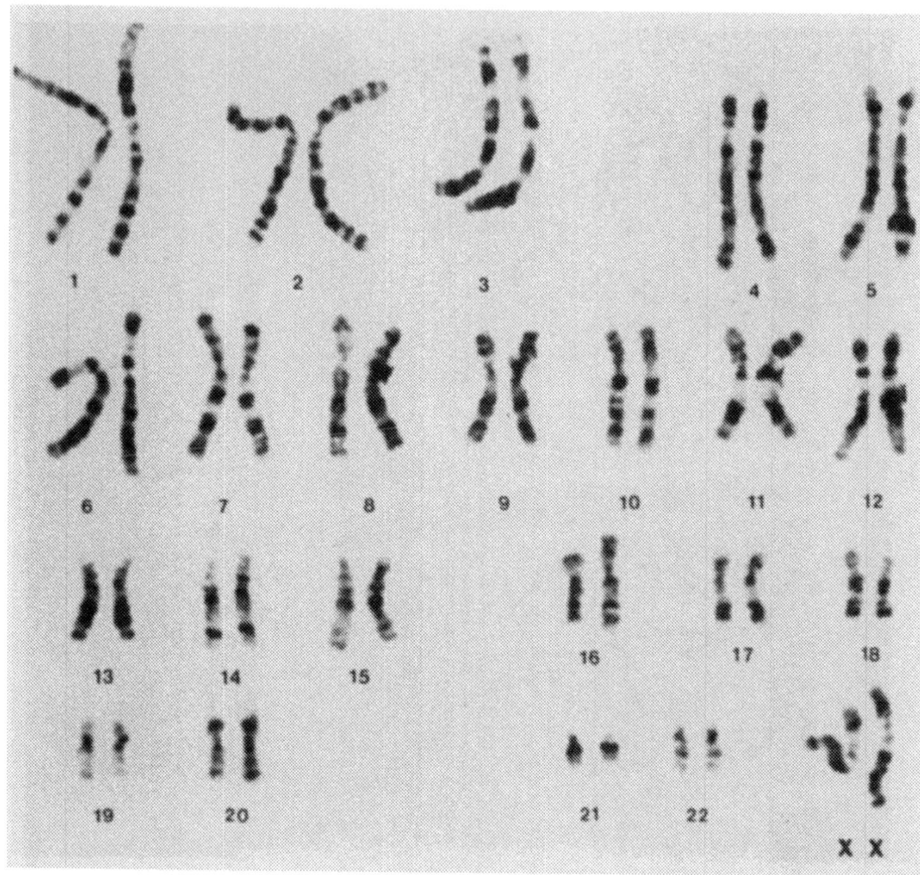

FIGURE 3 Karyotype of human chromosomes. A Giemsa (stained) banded normal female karyotype from amniocytes obtained for prenatal diagnosis. (Courtesy of Dr. Shivanand R. Patil.)

Figure 4 is the karyotype of a male cell and the only difference is in the sex chromosome, which is YX (pair number 23). The other 22 pairs of chromosomes are *autosomes*. They are essentially the same in both sexes—different, of course genetically.

Meiosis

To prevent a massive and fatal build-up of chromosomes as cell division proceeds, a *meiosis* or *reduction division* functions in the reproductive process. In the reproductive cells of the ovary or testis, each chromosome lines up opposite its partner. The rows of chromosomes then separate and one set, that is 23 single chromosomes, is eliminated from the cell. The separation takes place at the centromeres. Informative diagrams are presented in numerous biology texts (e.g., 5). As the female sex chromosome is XX, the remaining chromosomes and the ones expelled are identical. However, in the male sperm, the sex chromosome being XY means that either an X or a Y is eliminated. Here the dice game begins. Also, on a probability basis, approximately half of the sperm will be Y or X. If an X sperm fertilizes the egg, the fetus will be female; with a Y sperm it will be male; the sex of the fetus is determined by the male sperm.

Mitosis

After the sperm enters the egg, the 23 chromosomes in the egg and the 23 in the sperm gradually merge to form the new DNA. The "double chromosomes" then line up and divide. An equal number of singles (46 in humans) travel to

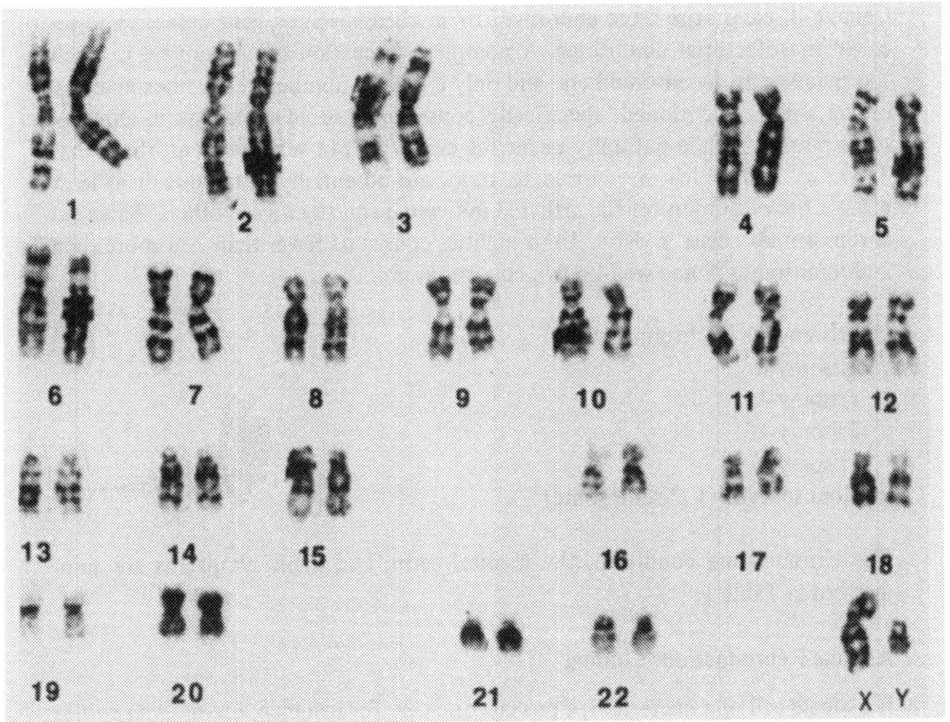

FIGURE 4 Karyotype of human chromosomes. A Giesma (stained) banded normal male karyotype from amniocytes obtained for prenatal diagnosis. (Courtesy of Dr. Shivanand R. Patil.)

opposite locations in the cell and the cell divides. This cell replication is called *mitosis* and it proceeds until the body is formed, and continues throughout the life of the organism. A highly illustrative sequence is pictured in Figure 5 (6).

Sex Identification

The sex of an individual organism can be established by identifying sex chromosome markers under a microscope. Again cheek cells are convenient for the procedure. When a *female cell* is treated on a glass slide with a conventional cell stain, a dark body can be seen on the inside wall of the cell membrane. This is the *female sex chromatin* or the *Barr body* (Dr. Mary Lyon of England identified it as the inactivated X chromosome) (Fig. 6a). Cells from a male organism, when treated with a fluorescent dye, show a "flashing Y," a highly fluorescent spot which is the *male sex chromatin* (Fig. 6b).

Genetic Diseases

Genetic diseases arise from abnormalities in chromosomes, gene defects, and so-called multifactorial conditions. A complete discussion of the subject is much too involved to be covered here, and only a limited number of chromosomal disorders will be mentioned; specifically conditions caused by errors in chromosome number. While naturally occurring errors are not very frequent, those that do occur, and which are carried to term, are potentially disastrous or at least place a heavy burden on the afflicted individual and significant others. Balanced chromosomes occur in pairs. Abnormalities consist of fewer than 2 or more than 2 chromosomes. Some well-known conditions are:

Trisomy-21 (3 chromosomes, Fig. 7)
Trisomy-8
Trisomy-13
Trisomy-18
Trisomy-X
Monosomy-X (1 chromosome)

The chromosome condition, the medical term, and main symptoms are summarized in Table 1.

Asexual Reproduction: Cloning

If a single cell of a bacterium, a prokaryotic cell, for instance, *Escherichia coli*, is placed on a nutrient agar medium, "bacterial growth" ensues by cell mitosis. A multitude of cells are produced by progressive cell division and all cells formed are identical to the original cell, that is, they are *clones*, descended from a single ancestor. The mass of cells does not represent a cohesive organism, rather each one is a separate organism. According to this definition of cloning, all of the cells of a mammalian organism are in essence clones of the original totipotent cell created by sexual reproduction.

Cloning attempts with egg transplants made with vertebrates, notably frogs, have achieved partial success (parthogenesis). However, the potential cloning of mammals, including humans, must at present and for some time to come, be relegated to the realm of science fiction.

GENE THERAPY

Dr. Williamson gives an analytical overview of the many aspects of gene therapy which were discussed in depth at a recent symposium (7). The concept of gene therapy in essence covers body defects or malfunctions which are the result of a defective gene including, of course, chromosomal defects which were covered in outline form in the previous section.

There are numerous genetic diseases which result from the malfunctioning of a *single gene*, such as sickle-cell anemia, thalassemia, growth hormone deficiency, hemophilia, and phenylketonuria, to name a few. Obviously, if gene therapy could be developed to control such ailments, there would be a great demand for it. Dr. Williamson, however, states: "... can it be done? At the moment, the answer is *unequivocally no*." A further complication is that many other diseases are due to more than one malfunctioning gene, and that in some diseases, it has not yet been possible to pinpoint the defective gene or genes. Sadly, developments in genetic engineering manipulation of human genes in the human body are, if at all, feasible far in the future. The article is a most interesting and educational treatise and must be recommended for study.

Cancer and DNA

One may question the appropriateness of touching upon some of the most recent findings on causes of human cancer. Perhaps the justification is that the basic structure of genetic engineering is DNA. A "popular" type of presentation by Wingerson (8) begins with the statement: "The seeds of cancer are present in normal DNA." The account deals mainly with the work of Weinberg at MIT, which has culminated in isolating and identifying actual cancer genes. Not unexpectedly, a considerable number of people in cancer research have, at least, reservations, or completely doubt the reported results. In essence, however, it should not be surprising that DNA breakage and reassembly by whatever means it occurs, could result in the phenomenon of malignancy.

Along the same line a more scientific report (9) described chromosome damage on human renal cell carcinoma and presents a rather definitive karyotype

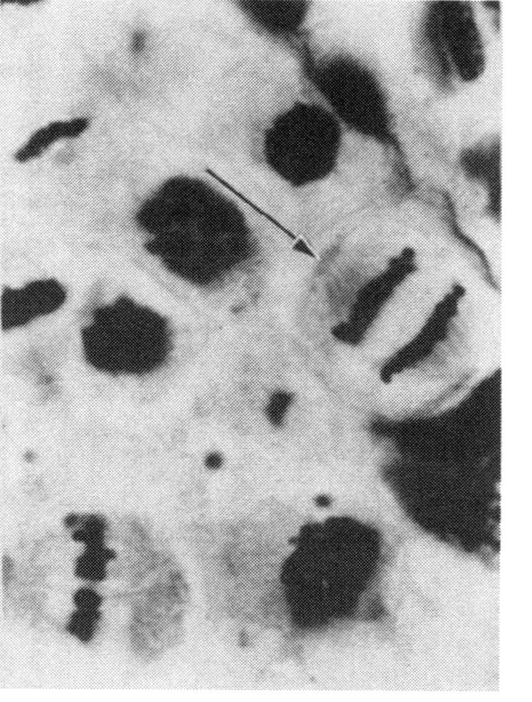

(a)

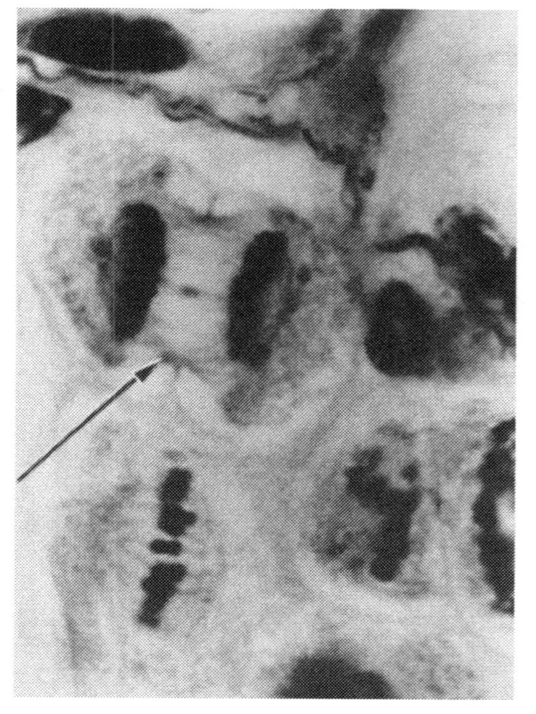

(b)

FIGURE 5 Sequences in mitosis. Photomicrograph of crayfish cell division (i.e. mitosis). Magnification 1000X. (a) Cell stage shows early anaphase. The chromosomes have already separated and begin to move apart along the spindle microtubules toward the centriole poles. (b) Late anaphase stage where chromosomes have moved to opposite poles.

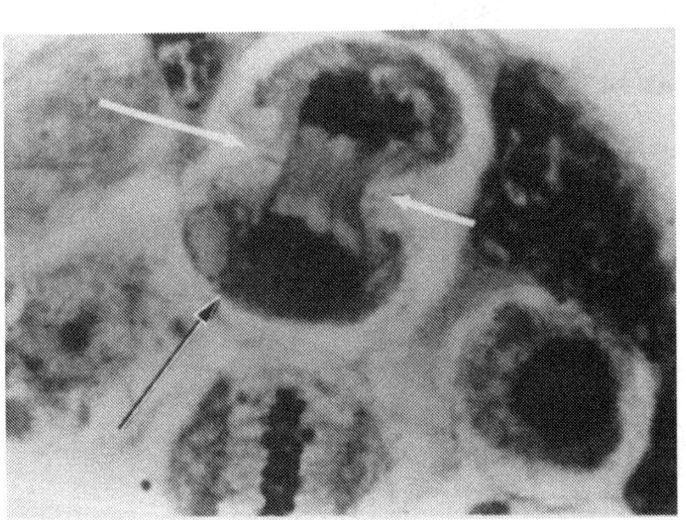

(c)

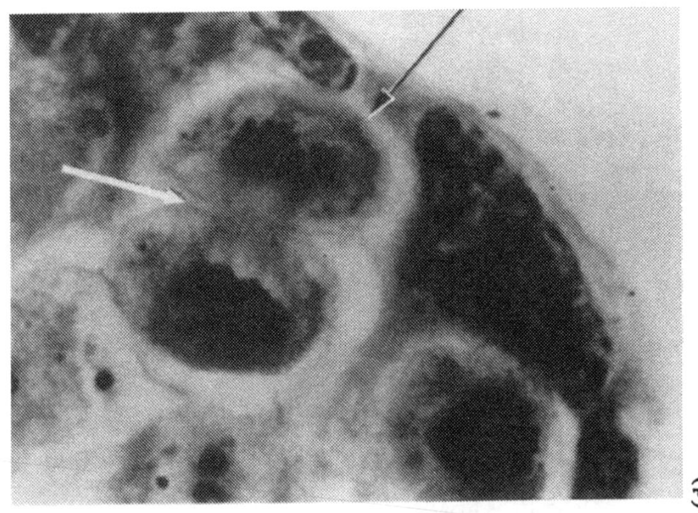

(d)

(c) Nuclei formation in progress in telophase stage. Note, indentation of cell wall begins. (——→)
(d) Late telophase stage represents completion of cell wall. Cell separation into 2 identical sister cells (clones) is imminent. [Courtesy D. Richard G. Kessel, Professor of Zoology, University of Iowa (Ref. 6).]

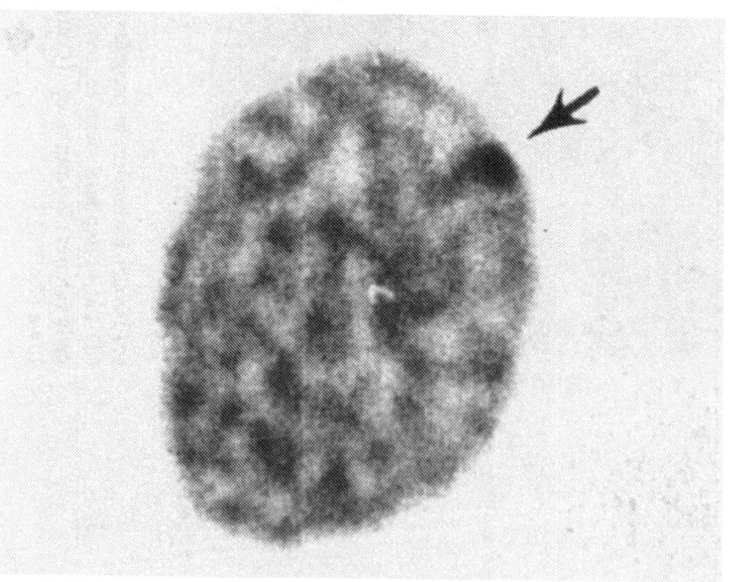
(a)

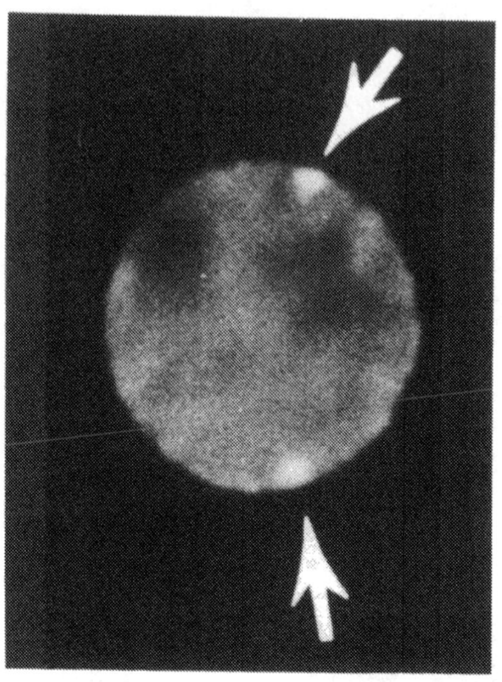
(b)

FIGURE 6 Sex determination. (a) Nucleus showing a darkly stained mass (Barr body). It represents an inactive X-chromosome and indicates that two X-chromosomes are present in this cell. (Courtesy of Dr. Shivanand R. Patil.) (b) A nucleus showing two brightly fluorescent spots (Y-chromatin masses) indicating that there are two Y chromosomes in this cell, an XYY condition. These spots represent the brightly fluorescent parts of the long arms of Y chromosomes when stained with fluorochrome such as Quinecrine mustard. (Courtesy of Dr. Shivanand R. Patil.)

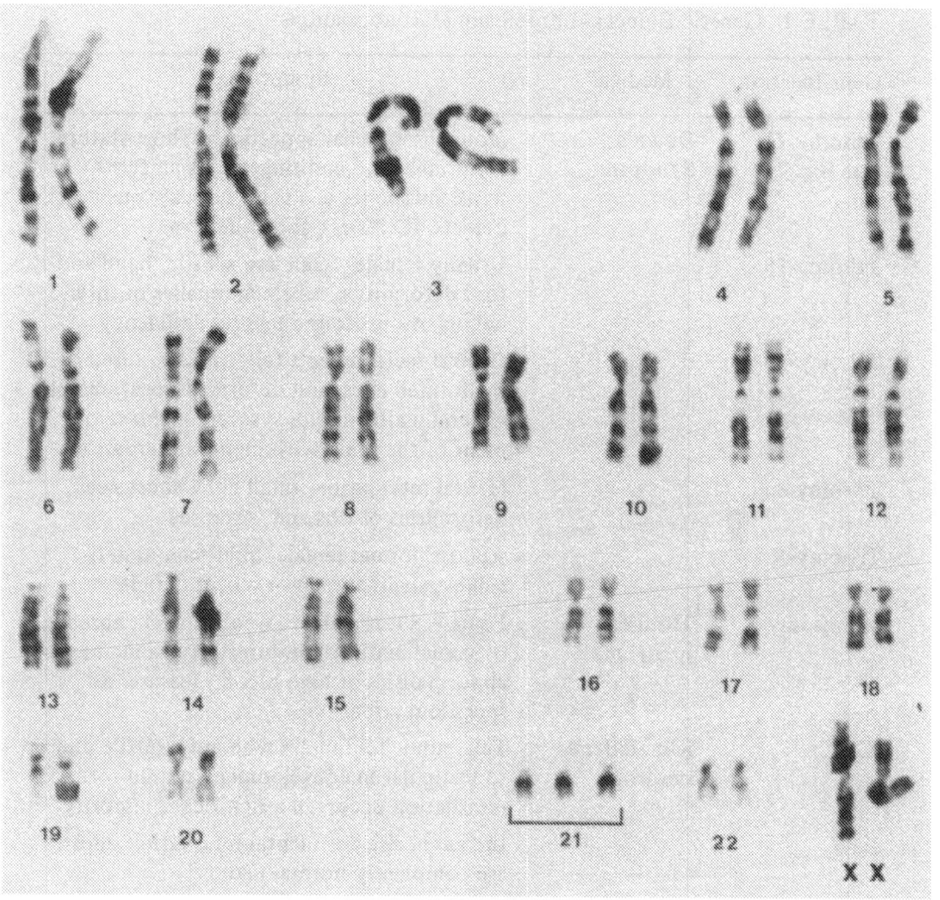

FIGURE 7 Karyotype of trisomy 21. A Giemsa-stained banded karyotype from amniocytes of a female Down syndrome fetus. Note the presence of three chromsome 21s. (Courtesy of Dr. Shivanand R. Patil.)

TABLE 1 Genetic Defects–Chromosomal Abnormalities

Gene location	Medical	Symptoms
Trisomy-21 (see Fig. 7)	Down's Syndrome	Mongolism, facial appearance, short stature, abnormal hand and fingerprint patterns, weak muscle tone, a third have serious heart defects, IQ 20–50, short lifespan.
Trisomy-18		Usually female, small and spastic, hand and foot deformities, severe anomalies of internal organs, profound mental deficiency
Trisomy-13		Central facial defects (cleft palate, fused or malformed eyes, and deformed nose), major visceral malformations, severe maldevelopment of the brain with mental retardation
Trisomy-8		Mental retardation, small chin, short neck, deformities of ribs and vertebrae
Trisomy-X		Almost normal female, mild mental deficiency, slight speech or motor deficits
Monosomy-X	Turner syndrome	Female, short stature, webbed neck, absence of sexual maturation, infertility, some heart abnormalities or high blood pressure, no specific retardation
XXY	Klinefelter's syndrome	Tall, eunuchoid males who are infertile due to testicular maldevelopment, a mild form of retardation occurs in a significant minority
XYY		Increased risk for mental retardation, most are completely normal males

showing the abnormal chromosomes. The essence of this study is not concerned with causative matters, but with factual observations of abnormalities in chromosome structures. A scholarly discussion by Dulbecco (10) covers the most up-to-date opinions regarding potential causes and their interpretation.

A most interesting report has been prepared by Miller (11), covering some of the most recent findings on initiation of cancer and plausible interpretations. The work of a number of research groups is summarized without going into specific details, while preserving the essential aspects of the studies. One finding would indicate that a single nucleotide substitution in what otherwise would be a normal sequence could result in a genetic change leading to cancer. A possible interpretation of cancer induction is that a normal cellular gene, a so-called cellular oncogene (or proto-oncogene) may become "activated" and be at the root of malignancies.

Premkumar et al. (12) report that the human bladder carcinoma oncogene shows only a *single point* of mutation. The change is that of guanosine into thymidine, resulting in the placing of valine for glycine in the protein chain.

Nonrecombinant Gene Transfer

A fascinating experiment in animal husbandry involving embryo manipulation was described by Bond (13). The following description is taken in toto from an Office of Naval Research publication.

> Scrambled Calves
>
> For perhaps the first time, calves have been produced that genetically have two fathers and two mothers. The experimental work was done by Steen Willadsen and Carol Fehilly at the Institute of Animal Physiology, Cambridge, UK. In the normal mammalian cloning procedure, an embryo is removed from a cow or sheep and divided into pieces; these pieces, when inserted into the uterus of a host mother, eventually result in genetically identical animals. The novelty in the recent Cambridge demonstration was the employment of two pairs of parents during early embryonic fusion. This innovation was achieved as follows:
>
> 1. A Frisian cow and a Hereford cow were artificially inseminated with semen from a bull of the same breed.
> 2. The two embryos were removed from the cows after a few cell divisions, when they consisted of 20 to 30 cells.
> 3. Each embryo was divided into four parts. These parts were put together in a culture, where they fused to form four mixed embryos.
> 4. The four mixed embryos were implanted, in pairs, into two host cows.
>
> Four calves were born; one was pure Frisian, one was pure Hereford, and the other two contained cells from all four parents.
>
> The two "chimerical" calves provide a most interesting genetic outcome. They are not crosses, in which there are combined some genes from both breeds in every cell of their bodies; instead, every cell in their bodies is either pure Frisian or pure Hereford. According to the investigators, some of the most obvious gross manifestations are in the skin, where tiny patches of Frisian and Hereford distinctive color may be observed.
>
> Quintuplet lamb clones also have been produced in the Cambridge laboratory. The experiments may lead to the successful growing of finer embryonic

fragments; with present methods, division into more than four viable fragments is extremely difficult.

REFERENCES

1. Mulligan RC, Berg P. Expression of a bacterial gene in mammalian cells. *Science* 209(4463):1422–1427 (1980).
2. Anderson WF, Diacumakos EG. Genetic engineering in mammalian cells. *Sci. Am.* 245 (1):106–121 (1981).
3. Stewart TA Wagner EF, Mintz B. Human β-globin gene sequences injected into mouse eggs, retained in adults, and transmitted to progency. *Science* 217(4564):1046–1048 (1982).
4. Karp LE. (1976). *Genetic Engineering*. Nelson-Hall, Chicago.
5. Stent GS (1971). *Molecular Genetics*. W.H. Freeman & Co., San Francisco (1971); also Mazia D. How cells divide. *Sci. Am.* 205(3):100–120 (1961).
6. Beams HW, Kessel RG. Cytokinesis: A comparative study of cytoplasmic division in animal cells. *Am. Sci.* 64(3):279–290 (1976).
7. Williamson B. Gene therapy. *Nature* 298(5873):416–418 (1982).
8. Wingerson L. Found: cancer genes. *Discover* 3(6):76–79 (1982).
9. Pathak, S et al. Familial renal cell carcinoma with 3;11 chromosome translocation limited to tumor cells. *Science* 217(4563):939–941 (1982).
10. Dulbecco R. The nature of cancer. *Endeavour, New Series* 6(2):59–75 (1982).
11. Miller JA. Spelling out a cancer gene. *Sci. News* 122(20):316-317, 319 (1982).
12. Premkumar E et al. A point mutation is responsible for the acquisition of transforming properties by the T24 human bladder carcinoma oncogene. *Nature* 300(5888):149–152 (1982).
13. Bond NA. Scrambled Calves. *Eur. Sci. Notes*. Office of Naval Research, London, ESN 36–7 (1982) by permission of the Dept. of the Navy.

13
Precautions and Regulations

Concerns over potential hazards in recombinant alterations of viable entities have been expressed by members of the scientific community and the lay public. The history of developing regulations is already long, involved, and often confused by redundant and baseless fears and allegations. A well-conceived publication entitled *Genetic Technology*: *A New Frontier* was created by the Office of Technology Assessment (97th Congress) (1). Its purpose, a documentary of informational and statistical presentations of major aspects and their potential impacts, successfully blends scientific principles and descriptive writing. The book is of interest to the technologist and also to readers with limited scientific background. A quote from the conclusion in the report is worth citing:

> Thus far no *demonstrable* harm associated with genetic engineering, and particularly recombinant DNA, has been found. But although demonstrable harm is based on evidence that damage has occurred at one time or another, it does not mean that damage *cannot* occur.

This cautiously worded statement effectively portrays the present status. No trouble so far, but be careful!

The most recent federal directives concerning guidelines for recombinant research are dated August 27, 1982. Two pertinent documents are listed in Ref. 2. As these directives are rather detailed and voluminous, many institutions and companies have prepared summary documents which contain the most essential information in explicit form.

The problem of government control is, of course, worldwide. What is particularly interesting is that the regulations recently imposed by Japan are more stringent than those of any other country. A news item (3) reports that the only host vectors permitted are *Escherichia coli* K12, *Bacillus subtilis* Marburg 168 (the harmless soil bacterium), and *Saccharomyces cervisiae*, a yeast vector.

All areas of adjunct interest were explored in depth at a 1981 Battelle Conference. The proceedings were published in five volumes. The main subject categories are listed in Ref. 4.

The controversial problem of the "Academic-Industrial Complex" was analyzed by Culliton (5). The accelerating growth of industrial investment in academic research raises the vexing question of how much control might be exercised by industrial research interests at the expense of basic academic research. The investments are in the multimillion dollar category, and thus very tempting.

Many academic scientists will find it difficult to resist such opportunities. Most such arrangements are, not surprisingly, in the genetic engineering disciplines.

PATENT SITUATION

The previously mentioned patent of Cohen and Boyer covering recombinant techniques (6,7) is a process patent. The licensing features were also cited in Chapter 11. At present this patent appears to be completely valid and functional. The issue of a second U.S. patent to Cohen and Boyer signifies that the inventors have materially strengthened their patent position.

This patent (No. 4,468,464) is a composition of matter patent, which covers "Method and compositions–for replication and expression of exogenous genes in micro organisms." This means that the use of a plasmid which has been created by the recombinant technique and which possesses certain characteristics and is also capable of being cloned, is protected under the patent.

As of November 1982, some complications seem to be developing (8). A claim of coinventorship has been made, which will cloud the issue significantly.

ECONOMIC ASPECTS

In keeping with the times, the journal *Nature* has started a "Nature guide to biorіches." The first such guide was published in the August 12, 1982 issue (p. 599).

As stated:

> Nature will in future publish a monthly listing of the stock performance of 15 representative biotechnology companies in the United States.

The "scoreboard" of performance will appear in the second issue of each month. Although several hundred companies have sprung up, the consensus of the financial community is that most of them will wash out.

DOCUMENTATION

The need for a comprehensive program of literature retrieval was stressed in the Introduction. Another vital source of information is the availability of a database of DNA sequences. This matter is now receiving attetnion. The National Institutes of Health have awarded a working contract to an industrial concern to collate and distribute DNA database information (9). Also, such information is already available from the European Molecular Biology Laboratory at the University of Heidelberg in Germany.

SUPERVISION CONCERNS

After a period of relative stability in recombinant supervisions, some new concerns have surfaced. The situation is well presented in condensed form in *Science News* (10). A government panel recommended the creation of an "oversight body" to monitor any potential "human genetic engineering" activities.

The concern is possibly highlighted by the recent announcement of the creation of supersized mice. Palmiter et al. (11) describe the fusion of a DNA fragment containing a mouse gene to the structure gene of rat growth hormone and microinjection of the recombinant sequence into fertilized mouse eggs. While this constitutes an important research accomplishment, with significance in growth hormone functioning, and possibilities of genetic defect correction, one cannot escape the thought: "supersized mice—that's all we need!"

EPILOGUE

This treatise was written at a time when the subject of genetic engineering was in a tumultuous phase of development. While the soundly developed principles are currently valid and while many of the reported findings on recombinant procedures are well established, there is much uncertainty about the scale of future growth to be expected. Also, scaling-up of procedures and commercialization of interesting and promising products still are beset with doubts, and progress in these areas may face many pitfalls. In the final analyses, however, genetic engineering is such a powerful tool, and the potential benefits derived from it are so attractive and promising, it is a safe conclusion that its momentum will carry it to incredible heights.

REFERENCES

1. *Genetic Technology: A New Frontier.* (1982). Office of Technology Assessment, Congressional Board of the 97th Congress, OTA Publishing Staff. Westview Press, Boulder, CO.
2. Recombinant DNA research, actions under guidelines. *Fed. Reg.* 46 (126): 34454–34460 (1981). Dept. of Health and Human Services, National Institutes of Health; also: Guidelines for research involving recombinant DNA molecules. *Fed. Reg.* 47 (167): 38048–38068 (1982).
3. Anonymous, *Chem. Week* 131 (5): 27 (1982).
4. *Proceedings of the 1981 Battelle Conference on Genetic Engineering*, Vols. I–V, Keenberg, M (Ed.), published by Battelle Seminars and Studies Program.
5. Culliton, BJ. The academic-industrial complex. *Science* 216 (4549): 960–962 (1982).
6. Cohen, SN, Boyer, HW. U. S. Patents 4,237,224 and 4,468,464 assigned to Leland Stanford University, Process for Producing Biologically Functional Molecular Chimeras, issued Dec. 2, 1980.
7. Anonymous. *Chem. Eng. News* 60 (33):7 (1982).
8. Budiansky, S. Biotechnology patent challenged. *Nature* 300(5890): 303 (1982).
9. Lewin, R. Long-awaited decision on DNA database. *Science* 217(4562): 817–818 (1982).
10. Anonymous. Ethics for gene splicers. *Sci. News* 122(23): 361 (1982).
11. Palmiter RD et al. Dramatic growth of mice that defvelop from eggs microinjected with metallothionetin-growth hormone fusion genes. *Nature* 300 (5893): 611–615 (1982).

14
Update

As stated in earlier chapters, the activities expected and actually accomplished are numerous and exciting. The almost feverish preoccupation in utilizing the recombinant technique accelerated in the late 1970s, reached a peak in early 1980, and then settled down to a more realistic and steady pace. The inherently difficult character of the procedures required some backtracking and in-depth investigation. At present research and applications development are well organized and progress is steady and productive.

To describe recent accomplishments in a reasonably condensed manner is difficult and would require a booksize presentation. Thus, I must use my own judgment to present a highly condensed summary.

REVIEW TYPE PUBLICATIONS

An overview of genetic engineering developments, as well as accomplishments, is given by Weaver (1) in a scientific-popular style. Combining historical sequence and spot highlights, the treatise encompasses DNA fundamentals such as chemical composition and behavior–replication and expression–as well as cutting, insertion, and recombination. Successes resulting from the recombinant technique in animal husbandry, plant improvement, and new pharmaceuticals are illustrated, and progress in anticancer techniques are described. The text material is greatly enhanced by the presentation of a large number of color illustrations by Ted Spiegel.

A fairly extensive treatise of biotechnology matters was published by Bjurstrom (2). Although it deals mainly with fermentation technology and associated equipment, it contains information which is pertinent to the cloning operations.

Koplove (3) prepared a comprehensive compilation of biotechnology texts. It covers the whole range of the field and thus constitutes a worthwhile reference material. Two additional recent texts, not mentioned in Koplove's survey, are by Rodriguez and Tait (4) on recombinant DNA techniques, and Drlica's (5) work on DNA and gene cloning. Both are highly informative and explain some of the more detailed ramifications of the biochemical aspects.

Many of the pertinent journals publish yearly reviews of progress in science with emphasis on bioscience. Good examples are the editorail review of 1984 in *Discover* (6) and a more specific article on biochemistry and molecular biology in the German periodical *Nachrichten* (7).

Frequently, publications are sponsored by industrial concerns, or by trade associations which are of considerable interest and worthy of attention. Usually these can be acquired at no cost. For instance a pamphlet from the *Anti-Aging News* (8) describes free radicals in DNA damage associated with aging. A bulletin representing a practical guide on affinity chromatography is available from the French firm Reactifs IBF (9). Also a fairly extensive glossary of biotechnology terms is available from the Provesta Corp. (10). There are, of course, many other items which can be gleaned from trade publications and journals advertisements.

CURRENT LITERATURE

The American journal *Science* and the British publication *Nature* deserve continued attention. *Scientific American* often publishes major articles dealing with genetic engineering and related biological areas. *Science News*, and in particular, the *New Scientist* (British weekly) give excellent overall coverage with good attention to pertinent references of original reports. Much of current activities can be gleaned by perusing the advertisements in biojournals.

In recent years some journals have published one or more special issues devoted to biotechnology (e.g., February and November 1983 of *Science*; August 13, 1984 issue of *Chemical & Engineering News*). Actually, review articles are usually more helpful to the technologist than the detailed accounts of the original research papers. A particularly useful review is an editorial article in the May 1985 issue of *BioScience*.

A new source of informational literature is offered by Omec International. Recent publications cover a geographical directory of persons active in the field, the availability of federal funding sources, and a biotechnology patent digest (11).

A thorough review of developments in biotechnology internationally was prepared by Zomely-Neurath (12). Trends in biotechnology, including research on immobilized cells and enzymes, applied genetics, and pharmaceuticals processing are covered at length.

Documentation

The status of documentation in 1982, exemplified by the creation of databases, was cited briefly in Chapter 13. The utility of information retrieval on a worldwide basis has been expanded materially by the establishment of new bases and steps to integrate the system. The problems involved in this effort were summarized by Lesk (13). Formation of a task group on coordination of acquisition, distribution, and utilization should promote international functioning. Summary articles by Zass (14) on the variety of coverage in databases, and by Sprinzl (15) on nucleic acid data banks, are helpful sources. Most recently, a data bank has been established in Australia (16).

BASIC STUDIES

The composition and behavior of DNA and RNA, as well as the principles of microbiology seem to be well established. Still there are areas which are hazy and not well understood.

A recent concept is that of the Prion (proteinaceious infection particle). It is believed to be a virus-type entity, reportedly 100 times smaller than the smallest known virus, and possibly a protein without containing a nucleic acid. However, this concept is quite controversial, as it implies an ability to replicate without a DNA and RNA directive and replication is protein directed. This behavior would be in direct contrast to the basic dogma that genetic information always flows from nucleic acids to proteins. It is not surprising, therefore, that a great deal of activity is rampant on this subject.

Two recent reviews present the status of Prion research and its controversial aspects. Thus, Prusiner (17) gives an estimated treatment (10 pages) in *Scientific American* and Masters (18) published a condensed version. Both discuss the possible involvement in the devastating Scrapie disease of sheep and goats and in several human degenerative diseases.

The gene-splicing area underwent a thorough critique at the 1985 Asilomar (California) conference (19), particularly in relation to the conference held a decade ago. A main concern was further constraints which might be imposed on gene-splicing activities and ensuing applications. In spite of reservations concerning potential hazards, the conferees found that tremendous progress had been made, even with the existing precautional limitations. The events at the conference are conveniently summarized by Fox (20).

Membrane Action

All organisms are comprised of cells which are packaged inside a membrane, the cell wall. Within the cell, membranes may act in a compartmental manner, sealing-off certain areas for specific functions in the life cycle of the organism. The activity of membranes as a co-worker in the art of living goes well beyond the "bagging" assignment. Active membrane processes are involved in the regulation of solute transfer, and in turn by osmosis and ultrafiltration. Thus, the intricate and interrelative effects involved in membrane processes of all types has been recognized as a separate discipline, that is: membrane mimetic chemistry and this topic is treated in great detail by Fendler (21). This text should be consulted if the reader's problem involves consideration of membrane questions.

Genetic Code

A surprising recent finding is the apparent divergence of the genetic code in protozoans and possibly in bacteria. The code, signifying amino acid sequence in proteins as dictated by triplets of bases in mRNA has been considered a universal fact. However, Caron and Meyer (22) and Preer et al. (23) have now established that the triplets UAA and UAG, which in "the code" mark termination of the translating sequence into protein, actually code for glutamine in the lower life forms, specifically in ciliated protozoans. The situation will undoubtedly call for extended research in the future. The present status is nicely presented by Fox (24).

DNA Synthesis

An instructive discussion of the options available for DNA synthesis was published by Bremner et al. (25). The review covers two basic methods practiced as manual, semiautomatic, and fully automated syntheses, with reference to instrument design and importance of operating features.

Z-DNA

The existence of Z-DNA, the left-handed configuration of DNA, is covered briefly in Chapter 3. This matter receives continued attention. Kolata (26) presents an overview which stresses the role of Z-DNA in regulatory processes. He states that Z-DNA is not simply a left-handed version of the well-known right-handed B form, but that it has a different conformation. Additionally, a brief note in *Bioscience* (27) calls attention to the studies of Hall et al. (28), who report the production of left-handed Z-RNA.

Split Genes

The fascinating existence and behavior of split genes in higher organisms (e.g., *not* in *E. coli*, where genes are continuous), is thoroughly discussed by Danchin and Slonimski (29). Split gene segments were a surprise discovery in 1977, and their significance and the matter of management of the segments by the organism have been and still are beset with uncertainties. The treatise covers the latest concepts and ongoing research, with lucid diagrams to illustrate possible mechanisms of intron excision and gene reorientation.

Gene Probe

A fairly new development is the commercial production of DNA probes to detect microbial DNA. The basic idea is to clone a piece of DNA that is unique to a particular bacterium or virus, and to use this cloned DNA as a probe in DNA-DNA hybridization to detect the presence of that microbial DNA in the clinical specimen. There have been several reports of using this technique (particularly in *Clinical Microbiology*) with ^{32}P as the labelling agent. The development of nonradioactive-labeling techniques, such as the biotin–avidin system available from ENZO (30), will make this a practical system for the clinical lab if it is as sensitive as the ^{32}P method. Another commercial DNA probe is marketed by Gene-Probe (31).

PLANT GENETICS

In Chapter 9 the statement was made that Ti plasmids are effective only if used with dicotyledonous plants. Very recently Hooykaas-Van Slogteren et al. (32) were able to show that the Ti plasmid DNA (T–DNA) could also be integrated into monocotyledons. This is an important finding as many food crops are monocots. The T-DNA transfer to the plant cells was accomplished by using strains with a wild-type Ti plasmid.

Protoplast fusion, a method used to create novel hybrids by inducing two different plant cells to fuse (see Chap. 9) has apparently not been as successful as it was expected to be. The difficulties encountered in rather intensive research over a decade were analyzed by Burgess (33). The hope was to beneficiate crop plants, such as soya beans and cereal species. But soya beans failed to regenerate protoplasts, sugar beet protoplasts formed cell walls but did not divide, and all cereal species investigated resisted fusion or regeneration. As the need to improve crop plants is a high priority, it is reasonable to expect that such efforts will be continued in spite of the present stalemate.

Transposable genetic elements in *Maize*, mobile genes discovered by McClinton (see Chap. 9), form the subject of a comprehensive report by Fedoroff (34). It is concluded that such elements may be of significance in evolution and that their properties might be used to affect the structure of genes and genomes.

Molecular genetics is the subject of a recent book by Downey et al. (35) (1983), which covers the plant aspects in depth and thus should be a valuable reference text.

MEDICINES AND MEDICAL APPLICATIONS

Health concern is an overriding aspect of human and animal life. Not surprisingly then, publications of advances in the area are digested with great empathy. Also, the rate of activity in research is highly stimulated by successes and lead to further intensification of efforts. One must realize that the vast number of accomplishments results in a proliferation of research reports and news items, which in many cases are of transitory interest and often are soon superceded by newer findings. Therefore the only way to keep abreast of progress is to consistently screen appropriate journals in the field.

This situation makes it difficult to decide which new findings should be cited here and to what extent they should be covered. Essentially there are two distinct areas. One is the creation of new or improved medications and the other area comprises the applications of techniques to medical procedures.

To maintain reasonable brevity, choices have to be made. Hopefully, the items to be covered will reflect highlights; but one person's conception of highlights might be another person's idea of a valley.

Medicines

The situation concerning the medical use of genetically engineered products is in a state of flux. This is so because the testing periods, for human use in particular, are long and costly. The verification of drug effectiveness is often difficult to establish and the problem of potential side effects can be perplexing.

It is significant that a major article was published in a clinically oriented journal by Engleberg and Eisenstein (36). The discussion covers such subjects as the advantages of cloned products, visualization of a cloning scheme, immunodiagnostic and therapeutic applications, as well as vaccine developments, use of cloned DNA as a diagnostic reagent, and products of cloned genes. This extensive overview, supported by 157 references, covers both medicines and applications to diagnosis.

A review directed more specifically to pharmaceuticals produced through genetic engineering was written by Vane and Cuatrecasas (37). These authors take much the same approach, as the "medicine" use and the "diagnostic" use overlap greatly in applications. The fact that a recombinantly produced compound is obtained in high purity, compared to purification from animal sources, is of course stressed as a significant advantage. Further discussion then deals with specifics, such as human insulin, growth hormones, monoclonal antibodies, and vaccines, followed by status of chemical synthesis, problems of drug delivery, and future competition between cloned substances and synthetically produced chemicals.

The recombinant techniques are being extended to the creation of medicinal chemicals. For instance, Hopwood et al. (38) describe the production of "*hybrid*" *antibiotics*. This appears to be the first report on this subject. The authors utilized recent developments in molecular cloning systems applied to *Streptomyces* which permit the isolation of biosynthetic genes for a number of antibiotics using bacterial plasmids. Novel compounds were obtained by gene transfer between strains producing the antibiotics actinorhodin, granaticin, and medermycin.

Insulin

The commercial synthesis of insulin is described in detail in Chapter 10. It is now well accepted and the product humulin is being marketed worldwide. So far humulin seems to be the only medicinal product to reach the market stage.

Interferons

There is considerable activity in the investigation of interferons. Many trials in experimental chemical uses are underway. It is still an up-and-down situation and it appears that the interferons may find profitable use on a specificity basis.

In the area of viral disease, interferon has proved to be an effective antiviral agent. But in the hoped-for anticancer field it has not been established as being uniformly helpful. Some successes have been claimed, but most trials have been essentially inconclusive. The status of interferon in viral and cancer studies have been discussed at various stages of development. Thus, Malpern (39) in a brief note in 1983 expresses doubts over its utility. But similar later publications sound more hopeful. Clemons (40) published a note on interferons and oncogenes (cancer/genes) where interferon is labeled as a potential cell growth inhibitor. Similarly, Balkwill (41) presents an overview of interferon's role in clinical trials. Again, some success has been experienced with cancer patients, especially with alpha interferon, but "interferons are not a miracle cure for cancer." Fortunately, the interferons" role (α, β, and γ) in viral diseases is more promising, particularly in treating the common "cold." As evidence that interferon will likely be a profitable commercial product, one should cite that in 1985, Schering-Plough Corp. announced that construction of facility for production of interferon located in Ireland was underway.

On the more basic side, a treatise by Pestka (42) covers the principles of gene isolation and cloning in *E. coli*. Helpful schematics of biochemical procedures and available information on gene sequences are shown.

Factor VIII

The blood clotting factor, a protein, has been successfully manufactured by recombinant techniques. Two good surveys contain the most helpful information on the intense activities for the technologist: Cherfas (43) and Brownlee and Rizza (44). Obviously, the principal task was that of characterizing the human factor VIII gene. This has been reported in detail by Gitschier et al. (45). The expression of the factor from recombinant DNA clones was reported by Wood et al. (46), and the structure of the polypeptide protein was delineated by Vehar et al. (47).

Also, the process of molecular cloning of a cDNA encoding factor VIII.C, a basic step in producing the human coagulation factor, has been reported by Toole et al. (48). A pertinent brief note by Beardsley (49) discusses the cooperative efforts of Genentech with two medical centers in England. Obviously, the success in actually manufacturing this protein compound is of enormous importance to those who suffer from hemophilia.

The use of *monoclonal antibodies*, complexed with chemotherapeutic agents is reminiscent of Erlich's "magic bullet" against syphilis in the early years of this century. Application of such complexes in testing blood for the presence of the HTLV-III virus, believed to be responsible for the acquired immune deficiency syndrome (AIDS) is being instituted on a large scale (50). Occasional news items indicate that complexes of chemotherapeutic agents with monoclonals are being investigated in anticancer treatments; this research seems to be in the beginning stage only.

Vaccines

Considerable hope is expressed in the possibility of creating synthetic vaccines through recombinant techniques. A major effort is underway for *foot and mouth disease*. The preparation of synthetic peptides which can be linked to a carrier protein is a promising approach, as described by Rowlands (51). Studies on hepatitis B vaccines are summarized by Williams (52). It may be possible to utilize a synthetic subunit rather than the complete viral constituent. Such a procedure would alleviate serious side reactions which are encountered with whole-virus vaccines.

Creating a vaccine against *Trypanosomiasis* (sleeping sickness) has a high priority, as this disease constitutes a devastating scourge of humans and animals alike. Though there is a faint hope that a vaccine might ultimately be prepared, the present state of knowledge makes it unlikely. The intricate behavior of the trypanosome, a unicellular parasitic protozoan, enables it to change the antigen of its surface coat in response to the mammalian immune system's reaction to the invader. Thus, while the immune reaction wipes out an existing infectious population, some of the trypanosomes have already acquired a new surface coat and proliferation begins anew. A very detailed and fascinating account of behaior was presented by Donelson and Turner (53). There, the basic concerns are well illustrated and placed in proper perspective to the overall problem of the disease and the specific research needed to hopefully lead to a solution.

Another important task is the creation of a vaccine against malaria, a disease that has plagued mankind for untold centuries. Despite large-scale efforts to eradicate it, it is still a continuing threat and widespread in many parts of the world. Young et al. (54) state that it "may be the most promising target for a vaccine by using a protein of the human malaria parasite." The *circumsporozoite* (CS) protein has been successfully expressed in *E. coli*, and in mice experiments, the recombinant products resulted in high titers of antibodies. In an extensive study of the CS protein, Ballou et al. (55) also have had success in preparing recombinant proteins to block malaria invasion at the sporozoite stage of the parasite.

As the basis of vaccine action rests on the concept of *immunology*, particularly the involvement of *B and T cells*, the matter of action and interaction of these cells is of utmost importance. Therefore, research on cellular immunology is an ongoing and vital activity. Howard (56) prepared an instructive overview of presently held concepts with particular emphasis on a newly proposed interaction mechanism. The mechanism evolved from a study by Lanzavecchia (57), who concludes that ". . . surface immunoglobulin (plays a role in) concentrating antigen on specific B cells and suggests that this may be the basis of clonal selection in the specific antibody response to low antigen concentrations."

Gene Therapy

Gene therapy is probably the most difficult and also the most controversial area which might benefit by genetic engineering. It involves intervention in the genome that carries a genetic defect. To do this in a mature organism is essentially impossible. Each cell of the organism can be presumed to contain the defect, and eradication of the defective gene in all of the existing cells would be a formidable task.

Obviously, procedures of an analytical nature to establish the presence of birth defects in the developing fetus are of basic importance. A relatively recent method to obtain specimens for cytogenetic examination is the chorionic villi sampling method, by which chorionic villi tissue which surrounds the fetal sac is collected. This is done by insertion of a catheter through the cervix instead of the presently prevailing amniocentesis procedure which requires needle collection of amniotic fluid. A brief note in *Science News* (58) summarizes the status of this technique.

In spite of the inherent difficulties associated with correction of defective genes in an established organism, there are reports of hopeful procedures. A note by Fox (59) describes experiments by Jaenisch at the University of Hamburg dealing with the insertion of genetic material into mouse embryos, with the help of a retrovirus. Chromosomal location of the viral gene has been established. While this basic work does not extend to gene therapy, it denotes success in transplanting a gene.

A more substantial review by Kolata (60) covers the research of Mulligan and co-workers at MIT. Earlier work by Mulligan et al. (MIT) showed that elimination of a component in the virus gene responsible for the virus's infective property, allowed the engineering of a noninfectious helper virus. This recombinant virus can enter a cell and insert its genes into the chromosomes. In another approach, these investigators have taken recourse to an "enhancer sequence" which causes expression of the gene after take-up by the cell. Also covered is work by Anderson and colleagues in cooperation with Yale University staff, on attempts to transfer beta-globulin genes to thalassemi mice. Additionally, efforts of some other groups are treated sketchily. A rather comprehensive paper by Anderson (61) is perhaps the most authoritative recent exposition on prospects for human gene therapy. In his conclusion, Anderson states "It now appears that effective delivery-expression systems are becoming available that will allow reasonable attempts at human gene therapy." The systems rely on treatment of bone marrow cells with a retroviral vector which contains the normal gene; a so-called somatic cell gene therapy technique. Appropriate to these accelerating efforts, the matters of safety and ethical concern are receiving continued attention. A summary of guideline revisions was published by Culliton (62) in May 1985.

Cancer Therapy

The intense activity in utilizing molecular biology findings for disease control in humans is emphasized by the most recent announcement that a patient's white blood cells when combined with interleukin-2 gave malignant tumor regression (62a). As customary, the experiments were conducted in animals, but some human trials are underway. The procedure is being readied for large-scale testing in selected human cancer patients.

In essence, the preparation of the medicinal injectable component uses the white blood cells from the patient's blood, which are then fortified with interleukin-2.

Interleukin-2 is described as a lymphokine (glycoprotein), a factor produced by helper T cells. The recombinant interleukin-2 was created by gene insertion into *E. coli*. Interleukin-2 is present only in small amounts in the human body and extraction from natural sources is difficult and produces insufficient amounts. Thus, the product made by cloning recombinant structures is the essential part for making the procedure a practical therapeutic method.

TECHNOLOGY

As the results of recombinant research are coming to fruition, development efforts in scaling-up procedures and in large-scale commercialization are getting increasing attention. In the previous discussion of design problems it was pointed out that a great deal of the experience in industrial biochemical operations can be utilized directly in the cloning stages of genetically engineered organisms. A cloning operation should be very much like a fermentation procedure. Consequently, a recent article by Swartz (63) is pertinent, as it treats the culturing of microorganisms in its overall aspects from inoculation to harvesting.

Commercialization requires that a process be economically sound. This factor is well treated by Guidoboni (64), who lists and discusses critical considerations in plant layout and equipment matching. A similar approach to the industrial aspect is taken by Michaels (65) in an analysis of problems incurring in the utilization of "modern biology." In essence, Michael's theme is for the need to acquaint technologists, in particular chemical engineers, with the language and techniques of the life scientists. A minimal degree of reeducation is a necessary requisite to participate in the new area. Also, Check (66) deals with the engineering challenge of biotechnology and states that "work with genetically altered organisms is causing fermentation engineers to reexamine their process designs and practices." On a more specific subject, Randerson (67) treats the production problems in generating *hybridomas* (see Chap. 12).

The development of manufacturing facilities is exemplifed by Damon Biotech's and Bio-Response's installations that will produce kilogram amounts of *monoclonal antibodies* (68). A summary note by Klausner (69) describes the blossoming industrial development in Ireland directed to biotech excellence. Specific mention is made of Schering-Plough's *interferon* facility.

With the proliferation of biotech activities, there is added concern of safety in all areas. This subject is analyzed in depth by Rhein (70), who presents important summaries of regulatory agencies and ancillary information.

The problem of "down-stream processing" in biological manufacturing is beset with an inordinate number of difficulties. This processing, for the most part, still involves the rupturing of the bacterial cells to isolate the desired component. Generally speaking, a genetically engineered bacterium, most frequently an *E. coli* cell, will produce a specific compound such as a protein, which is locked within the cell structure. Then the recovery is accomplished by cell rupture, followed by sequential extraction and purification steps.

Sharma (70a) recently reviewed the recovery steps required to isolate genetically engineered proteins from *E. coli* cloning operations. The basic problems are the solubilizing of the proteins which occur as precipitated matter within the denatured cell and the removal of the denaturing agent such that the optimal folding of the protein is preserved.

Patents

A proper perspective of the *gene-splicing patent* situation is difficult to assess. Conflicting reports are common and one's evaluation depends mainly on which account is more persuasive. The original and most farreaching Cohen–Boyer patent (Stanford University) was described in Chapter 10. So far it seems to hold up. A new patent was issued August 28, 1985, as reported by Norman (71). Another account by Anderson (72), however, highlights some controversial aspects which may be troublesome to resolve, as indicated by the Cetus Company's termination of its license agreement with Stanford (73).

A looming confrontation between Schering-Plough and Hoffman–La Roche over patents issued for manufacturing *alpha interferon* was apparently avoided by an intercompany agreement (74). The situation leading up to the potential conflict was documented by Beardsley (75). The status of interferon, particularly the potential applications and its commercial promise, is treated in a text by Panem (76), which is capably reviewed in *Nature* by Nelkin (77).

Recombinant research is accelerating and the proliferation of the patent activity is indicative of fruitful accomplishments. The impact of public policy and legal issues was treated at length by Adler (78) in a discussion of philosophy of patentable materials, the problems of international competition, research funding, and gene ownership. Some recent U. S. patents are summarized and a literature survey covering 73 references attests to the depth of treatment.

REFERENCES

1. Weaver RF. Changing life's genetic blueprint. *Natl. Geographic* 166(6): 818–847 (1984).
2. Bjurstrom E. Biotechnology. *Chem. Engg.* 92(4):126–158 (1985).
3. Koplove M. Books on biotechnology. *Chem. Engg.* 91 (23): 145–169 (1984).
4. Rodiguez RL, Tait RC. (1983). *Recombinant DNA Techniques–An Introduction.* Addison-Wesley, Reading, MA.
5. Drlica K. (1984). *Understanding DNA and Gene Cloning.* John Wiley & Sons, New York.
6. 1984–The year in science. *Discover* 6(1):60–77 (1985).
7. Biochemie und Molekulärbiologie 1984. *Nachrichten* 33(2):131–144 (1985).
8. Summerfield FW. Free radicals, DNA damage and aging. *Anti-Aging News* 5 (4):37–41 (April 1985).

9. Practical Guide for use in Affinity Chromatography and Related Techniques, 2nd ed. (1983). Reactifs IBF, Societe Chimique Pointet-Girard, Villeneuve-La-Garenne, France.
10. *The Language of Biotechnology–a Glossary*. (1984). Provesta Corp., Bartesville, OK.
11. Omec International, 1128 Sixteenth St., NW, Washington, DC 20036.
12. Zomely-Neurath CE. Third European Congress on Biotechnology. ONRL Report C-1-85, U. S. Office of Naval Research, London (Available from Box 39, FPO NY 09510).
13. Lesk AM. Coordination of sequence data. *Nature* 314(6009): 318–319 (1985).
14. Zass E. Verschiedene Datenbasen (Variety of Databases). *Nachrichten* 32 (7):578–580 (1984).
15. Sprinzl M. *Nachrichten* 32(7): 212–214 (1984).
16. Reisner AH, Bucholtz C. Australian databank. *Nature* 314(6009):310 (1985).
17. Prusiner SB. Prions. *Sci. Am.* 251 (4):50–59 (1984).
18. Masters C. Scrapie–perspectives on prions. *Nature* 314(6006):15–16 (1985).
19. Miller J. Lessons from Asilomar. *Sci. News* 127(8):122–23, 126 (1985).
20. Fox JL. Asilomar Anniversary Appeal. *Chem. Ind.* 1985(7):208 (1985).
21. Fendler JH. (1982). *Membrane Mimetic Chemistry*. Wiley-Interscience, New York.
22. Caron F, Meyer E. Does *Paramecium primaurelia* use a different genetic code in its macronucleus? *Nature* 314(6007):185–188 (1985).
23. Preer JR et al. Deviation from the universal code shown by gene for surface protein 51A in *Paramecium*. *Nature* 314 (6007):188–190 (1985).
24. Fox TD. Divergent genetic codes in protozoans and a bacterium. *Nature* 314 (6007):132–133 (1985).
25. Bremner J, Hamill B, Mackie H. Options for DNA-synthesis. *Am. Biotechn. Lab.* 3(2): 46–53 (1985).
26. Kolata G. Z–DNA moves toward real biology. *Science* 222(4623): 495–496 (1983).
27. *BioScience* 35(2): 134 (1985).
28. Hall K et al. Z–RNA–A left-handed RNA double helix. *Nature* 311(5986): 584–586 (1984).
29. Danchin A, Slonimski PP. Split genes. *Endeavor* 9(1): 18–27 (1985).
30. Enzo Biochem, Inc., 325 Hudson St., New York, NY 10013.
31. GEN-PROBE, 9620 Chesapeake Drive, San Diego, CA 92123.
32. Hooykaas-Van Slogteren GMS et al. Expression of Ti plasmid genes in monocotyledonous plants infected with *Agrobacterium tumefaciens*. *Nature* 311(5988):763–764 (1984).
33. Burgess J. The revolution that failed. *New Scientist* 104(1428):26–29 (1984).
34. Fedoroff NV. Transposable genetic elements in maize. *Sci. Am.* *250*(6):85–98 (1984

35. Downey K et al. (1983). *Advances in Gene Technology, Molecular Genetics of Plants and Animals*. Academic Press, New York,
36. Engleberg NC, Eisenstein BI. The impact of new cloning techniques on the diagnosis and treatment of infectious diseases. *New Eng. J. Med.* 311(14): 892-901 (1984).
37. Vane J, Cuatrecasas P. Genetic engineering and pharmaceuticals. *Nature* 312(5992): 303–305 (1984).
38. Hopwood DA et al. "Production of 'hybrid' antibiotics by genetic engineering. *Nature* 314(6012): 642–644 (1985).
39. Malpern J. Further doubt cast over interferon's future. *Chem. Ind.* 1983 (19): 725 (1983).
40. Clemons M. Interferons and oncogenes. *Nature* 313(6003): 531–532 (1985).
41. Balkwill F. Interferons: from common colds to cnacer. *New Scientist* 104 (1447): 26–28 (1985).
42. Pestka S. The purification and manufacture of human interferons. *Sci. Am.* 247 (2): 37–43 (1983).
43. Cherfas J. The engineering of a blood-clotting factor. *New Scientist* 104 (1432):19–20 (1984).
44. Brownlee GG, Rizza C. Clotting factor VIII cloned. *Nature* 312(5992): 307 (1984).
45. Gitschier J et al. Characterization of human factor VIII gene. *Nature* 312 (5992): 326–330 (1984).
46. Wood WI et al. Expression of active human factor VIII from recombinant DNA clones. *Nature* 312(5992): 330–337 (1984).
47. Vehar GA et al. Structure of human factor VIII. *Nature* 312(5992): 337–342 (1984).
48. Toole JJ et al. Molecular cloning of a cDNA encoding human antihaemophilic factor. *Nature* 312(5992): 342–347 (1984).
49. Beardsley T. Genentech claims factor VIII. *Nature, 309*(5963):3 (1984).
50. *Science* 225 (4667):1129 (1984).
51. Rowlands DJ. Synthetic peptides in foot and mouth disease vaccine research. *Endevour* 8(3):123–127 (1984).
52. Williams N. Building new vaccines. *Nature* 306(5942:427 (1983).
53. Donelson JE, Turner MJ. How the trypanosome changes its coat. *Sci. Am.* 252(2):44–51 (1985).
54. Young JF et al. Expression of *Plasmodium falciparum* circumsporozoite proteins in *E. coli* for potential use in a human malaria vaccine. *Science* 228 (4702):958–962 (1985).
55. Ballou WR et al. Immunogenicity of synthetic peptides from circumsporozoite protein of *Plasmodium falciparum*. *Science* 228(4702): 996–998 (1985).
56. Howard JC. Immunological help at last. *Nature*, 314(6011): 494–495 (1985).
57. Lanzavecchia A. Antigen-specific interaction between T and B cells. *Nature* 314(6011): 537–539 (1985).

58. Anonymous. New test for birth defects. *Science News* 124(8): 116 (1983).
59. Fox JL. Injected virus probes fetal development. *Science* 223(4643):1377 (1984).
60. Kolata G. Gene therapy method shows promise. *Science* 223(4643): 1376–1379 (1984).
61. Anderson WF. Prospects for human gene therapy. *Science* 226(4673):401–409 (1984).
62. Culliton BJ. Gene therapy guidelines revised. *Science* 228(4699):561–562 (1985).
62a. Rosenberg SA et al. *New Engl. J. Med.* 313(23): 1485–1492 (1985); Mark JL. a synopsis in *Science* 230(4732):1367–1368 (1985).
63. Swartz JR. The large scale culture of microorganisms. *Am. Biotech. Lab.* 3(1):37–46 (1985).
64. Guidoboni GE. Engineering for an economic fermentation. *Chem. Ind.* (British) 1984 (12): 439–443 (1984).
65. Michaels AS. Adapting mdoern biology to industrial practice. *Chem. Engg. Progr.* 80(6):19–25 (1984).
66. Check W. The engineering challenges biotechnology is posing. *Mosaic* 15 (4):20–27(1985).
67. Randerson D. Hybridoma technology and the process engineer. *Chem. Eng.* (British), 1984 (409): 12–15 (December 1984).
68. Anonymous. Two biofirms scale up monoclonals for industry. *Chem. Week* 143(13): 28–30 (1984).
69. Klausner A. Irish eyes shine on biotech. *Biotechnology* 2(11): 929 (1984).
70. Rhein R. A guide to biotechnology regulators. *Che. Week* 136(3): 26–34 (1985).
70a. Sharma SK. On the recovery of genetically engineered proteins from *Escherichia coli. Sepn. Sci. Technol.* 21(8): 701–726 (1986).
71. C. Norman, "Cohen-Boyer Patent Finally Issued," *Science 225*(4667), 1134 (1984).
72. Ian Anderson, "Gene Patent Granted: Now the Real Fight Begins, " *New Scientist, 103*(1420), 7 (1984).
73. Anon., "Gene-splice Patent Showdown," *Science News, 127*(20), 312 (1985).
74. Anon., "Interferon Patent Battle Avoided," *C & E News, 63*(20), 8 (1985).
75. T. Beardsley, "Contest in Prospect Over Interferon Rights," *Nature, 314* (6008), 207 (1985).
76. S. Panem, "The Interferon Crusade: Public Policy and Biomedical Dreams," The Brookings Institution (1984).
77. D. Nelkin, Bookreview, *ibid., Nature, 314*(6012), 647 (1985).
78. R. G. Adler, "Biotechnology as an Intellectual Property," *Science, 224* (4647), 357–363 (1984).

Appendices

APPENDIX 1 Frequently Used Reagents in Recombinant Procedures

Brij 35 poly-oxyethylene lauryl ether

Brij 58 polyoxyethylene-20 cetyl ether

Broth LB–aqueous solution:

bacto-tryptone	10 g/liter
bacto-yeast extract	5 g/liter
NaCl	5 g/liter

Broth M9–acqueous solution:

$Na_2HPO_4 \cdot H_2O$	6 g/liter
KH_2PO_4	3 g/liter
NaCl	0.5 g/liter
NH_4Cl	1 g/liter

EDTA ethylenediaminetetraacetic acid

EtBr ethidium bromide:

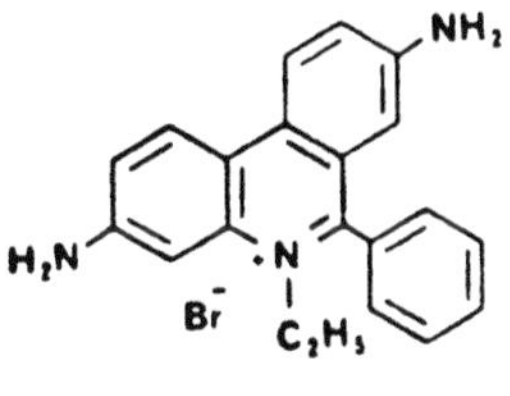

NH₂ H₂N N Br⁻ C₂H₅

Phosphate buffer:
di- and triphosphate mixture to pH ~ 6.9

TEN buffer:
10 m*M* Tris-HC1, pH 8.1
1 m*M* EDTA
300 m*M* NaCl

Tris–Tris (hydroxymethyl) aminomethane

APPENDIX 2 Some Common Abbreviations

PEG Polyethylene glycol

DOC Deoxycholate

HPLC High pressure liquid chromatography

Tris-HCl Tris (hydroxy methyl) aminomethane hydrochloride

EDTA Ethylenediaminetetraacetic acid.

HaeIII/ALU I digest Slash means *double* digest of DNA

ColEI imm. Immunity to Colicin EI

Phage λ imm. Immunity to Phage λ

trp E Tryptophan E gene

trp ED Tryptophan E and D gene

Amp Ampicillin

Cam or Cm Chloramphenicol

Ery Erythromycin

Kan Kanamycin

St Streptomycin

APPENDIX 3 Trade Publications

Genetic Engineering and Biotechnology Firms, U.S.A.-1981.

Sittig and Noyes, 84 Main St., P.O. Box 75, Kingston, NJ 08528. Quoted Price: $100

Genetic Engineering News, Monthly

Biotechnology Press Digest, Monthly, $185/yr

The Agricultural Genetics Report, 6 issues, $90

The Biotechnology Law Report, Monthly, $275/yr

All published by Mary Ann Liebert, Inc., 1651 Third Avenue, New York, NY 10128

APPENDIX 4 Bacterial Cloning Vehicles

H.U. Bernard and D.R. Helinski (1980). *Genetic Engineering*, Vol. 2. Plenum Press, pp. 133-167.

This publication represents a very complete compilation of cloning vehicles from *E. coli* and *S. aureus* and *Bacillus* with an extensive bibliography of 123 references.

The information for 28 *E. coli* plasmids covers:

Replicon
Size in kilobases
Selective markers
Single restriction sites
Comments on useful characteristics
Original references

E. coli plasmid cloning vehicles promoting the expression of inserted genes, Information given:

Promoter
Replicon
Cloning vehicle
Properties of cloning vehicle
Inserted gene
Plasmid carrying insert
Expression of inserted gene

The data for 12 *S. aureus* and *Bacillus* cover:

Natural occurrence
Used in
Size
Selective markers
Single restriction sites
Restriction sites available for cloning.

Glossary

Acidic Amino Acids Amino acids having a net negative charge at neutral pH.

Actinomycin D Antibiotic that blocks elongation of RNA and DNA chains.

Activating Enzyme (See Amino acyl tRNA Synthetase).

Active Site A point source in three-dimensional structure of proteins that is the catalytic center of the enzyme.

Adaptor Molecules Small RNA molecules (tRNA) that locate amino acids in their proper positions on an mRNA template during protein synthesis.

Adenosine Triphosphate (ATP) A high-energy phosphate ester which serves as the principal energy-storage compound of the cell.

Adenovirus Animal viruses which contain linear duplex DNA within an icosahedral protein shell.

Affinity Chromatography A technique of molecular separation in which molecules are attached to a solid and insoluble matrix such as a sepharose gel. Molecules which possess affinity to the bound molecule are retained and can be subsequently eluted.

Alleles The two members of a particular gene pair, located at corresponding points on homologous chromosomes.

Allergy Increased sensitivity to an antigen brought about by previous exposure.

Amino Acids (a) Organic compounds containing both amino, $-NH_2$ and carboxyl, $-COOH$ groups. (b) The building blocks of proteins. There are twenty common acids. All amino acids have the same basic structure, but differ in their side groups (R).

Amino Acid Sequence The linear order of the amino acids in a peptide or protein.

Amino Terminal The end of a polypeptide chain that has a free α-amino group.

Aminoacyl Adenylate An activated compound occurring as an intermediate in the formation of proteins.

Aminoacyl tRNA Synthetase Any one of at least 20 different enzymes that catalyze (a) the reaction of a specific amino acid with ATP to form enzyme bound aminoacyl-AMP (activated amino acids) and pyrophosphate and (b) the transfer of the activated amino acid to tRNA forming aminoacyl-tRNA and free AMP.

Amniocentesis Insertion of a needle into the amniotic cavity for the purpose of withdrawing a sample of the amniotic fluid.

Angstrom (Å) A unit of length equal to 10^{-8} cm.

Anticodon The three-base group on a tRNA molecule that recognizes and pairs with a three-base codon of mRNA, responsible to correctly position an amino acid into a growing protein chain.

Antigen Any foreign object that, upon injection into a vertebrate, will stimulate the production of neutralizing antibodies.

Autoradiography Detection of radioactive labels in molecular structures by exposure of films.

Autosome Any chromosome other than the sex chromosomes.

Bacterial Viruses Viruses that invade and multiply in bacteria.

Bacteriophage (phage) A virus that infects bacteria.

Barr Body The inactivated X chromosome. Also known as the female sex chromatin.

Base Analogs Purines and pyrimidines that differ slightly in structure from the normal nitrogenous bases.

Base-Pairing Rules The requirement that adenine must always form a base pair with thymine (or uracil) and guanine with cytosine, in a nucleic acid double helix.

Base Ratio The ratio of (A + T)/(C + G) in the DNA structure.

Basic Amino Acids Amino acids having a net positive charge at neutral pH.

β-Galactosidase An enzyme catalyzing the hydrolysis of lactose into glucose and galactose; in *E. coli*, the classic example of an inducible enzyme.

Blocked (Capped) 5′ Ends The 5′ ends of most eukaryotic mRNA's are post-transcriptionally modified by the addition of GTP in a 5′-5′ condensation.

Breakage and Reunion The crossing-over by physical breakage and crossways reunion of completed chromatids during meiosis.

Capsid A shell-like structure, composed of aggregated protein subunits, that encloses the nucleic acid component of viruses.

Capsomeres Protein subunits which aggregate to form the capsid.

Carboxyl Terminal The end of a polypeptide chain that has a free α-carboxyl group.

Cell The fundamental unit of life; the smallest body capable of independent reproduction. Cells are always surrounded by a membrane.

Cell Culture The "in vitro" growth of cells isolated from multicellular organisms.

Cell Cycle The timed sequence of events occurring in a cell in the period between mitotic divisions.

Cell-Free Extract A fluid containing most of the soluble molecules of a cell, after cell rupture.

Cell Fusion Formation of a single hybrid cell with nuclei and cytoplasm from different cells.

Central Dogma The basic relationship between DNA, RNA, and protein: DNA serves as a template for its own duplication and formation of mRNA and RNA in turn is the template in protein synthesis.

Centrioles Paired, cylindrical structures that lie just outside the nuclei of animal cells, involved in the formation of the spindle.

Centromere A region of the eukaryotic chromosome where it attaches to the kinetochore just before replicated chromosome separation, and so holds together the paired chromatids.

Chimera An individual composed of cells originally derived from two or more separated fertilized eggs. A modified plasmid or viral DNA.

Chloroplast A cell plastid (specialized organelle) in plants and algae containing chlorophyll pigments and functions in photosynthesis.

Chromatid The two daughter strands of a duplicated chromosome that are still joined by a single centromere.

Chromatin The nucleoprotein fibers of the eukaryotic chromosomes.

Chromosomal Tubules Spindle microtubules which originate at the kinetochores.

Chromosomes Threadlike structures which carry the hereditary material of cells and viruses.

Cilia Moving organelles located on the cell surface.

Cistron The genetic unit which carries information for the synthesis of a single polypeptide (enzyme or protein molecule); determined by the cis-trans complementation test.

Clone A group of cells all descended from a single common ancestor.

Cloning A sexual reproduction in which all the descendants making up a clone are derived from a single cell, and are genetically identical.

Coat Protein The external structural protein of a virus.

Code A directive for the production of a protein.

Codon A sequence of 3 nucleotides (in a nucleic acid) which codes for an amino acid or for the initiation or termination of a polypeptide chain.

Coenzymes Small molecules which associate with proteins to form active enzymes.

Colony A group of contiguous cells, usually derived from a single ancestor, growing on a solid matrix.

Complementary Base Sequences Polynucleotide sequences that are related by the base-pairing rules.

Complementary Structures Two structures which define each other, as for instance, the two strands of a DNA helix.

Concatemer Structure formed by the linear repeat of unit-size components.

CsCl Centrifugation (See Equilibrium Centrifugation.)

Cyclic AMP Adenosine monophosphate with a phosphodiester bond between 3′ and 5′ carbon atoms to form a cyclic molecule. Participates in regulating gene expression in bacterial and eukaryotic cells.

Cyclic GMP Guanosine monophosphate; same structure as AMP.

Dalton A unit of mass equal to the mass of a single hydrogen atom.

Degenerate Codons Two or more codons that code for the same amino acid.

Deletions Loss of a section or sections of the genetic material from a chromosome.

Denaturation The loss of the native configuration of a macromolecule. It can be caused by heat treatment, extreme pH changes, chemical treatment, or other denaturing agents. Loss of biological activity often results.

Density Gradient Centrifugation Molecular size separation through differential sedimentation. A preformed density gradient is established before centrifugation; fragments move to the density level where their density equals that of the solution.

Deoxynucleoside The condensation product of a purine or pyrimidine with the five-carbon sugar, 2-deoxyribose.

Deoxyribonucleic Acid (DNA) A polymer of deoxyribonucleotides linked by phosphodiester bonds. The genetic material of all cells.

Deoxyribonucleic Ligase An enzyme capable of joining together split fragments of DNA.

DNA Polymerase I The first bacterial enzyme found to catalyze the formation of the 3′-5′ phosphodiester bonds of DNA. DNA repair is most likely its chief biological function.

DNA Polymerase III Bacterial enzyme which catalyzes the formation of 3′-5′ phosphodiester bonds at a very rapid rate. Its chief role seems to be DNA replication.

DNA-RNA Hybrid A double helix that consists of one chain of DNA hydrogen-bonded to a chain of RNA by means of complementary base pairs.

Deoxyribonucleotide A compound which consists of a purine or pyrimidine base bonded to the sugar, 2-deoxyribose, which in turn is bound to a phosphate group.

Dimer Structure resulting from association of two monomer units.

Diploid State The chromosome state where each type of chromosome except the sex chromosomes is represented twice.

Disulfide Bond Covalent bond between two sulfur atoms in different amino acids of a protein.

Dominant An allele which exerts its phenotypic effect when present either in homozygous or heterozygous form.

Electron Microscopy A technique for visualizing material that uses beams of electrons instead of light rays and that permits greater magnification than is possible with an optical microscope. Resolutions of ~10 Å are attainable with biological materials.

Endonuclease An enzyme that makes internal cuts in DNA backbone chains.

Enzyme Protein molecules capable of catalyzing chemical reactions.

Episome In bacteria, genetic material in the cytoplasm separate and distinct from the chromosome.

Epithelium Tissue which acts as a covering or lining for any organ or organism.

Equilibrium Centrifugation A technique used to separate molecules by their density: used to separate nucleic fragments. When ultracentrifuged in a heavy salt solution that forms a density gradient, fragments move to the density level where their density equals that of the solution.

Established Cell Line Cultured cells of single origin capable of stable growth for many generations.

Eukaryote Organism with cells that have a membrane enclosed nucleus.

Eukaryote Cell The more complex of the two cell types. There is a membrane-enclosed nucleus, containing chromosomes, and the metabolic machinery is arranged in well-organized membrane-surrounded compartments.

Euploid Having the normal number of chromosomes for the species.

Exonuclease Enzyme that acts on the ends of DNA.

Female Sex Chromatin The inactivated X chromosome. Also known as the Barr body.

Fertilization Fusion of gametes of opposite sexes to produce diploid zygote.

Genome A complete set of genetic material.

Genotype The genetic constitution of an organism.

Glycoprotein Polypeptide to which sugar residues are attached.

Golgi Apparatus A complex series of flattened, parallel membranes which appear to function in granule formation and molecular processing. There lysosomes are formed and secretory products are packaged into vacuoles.

Growth Factor A specific substance that is needed to allow cell multiplication.

Hairpin Loops Regions of double helix formed by the pairing of two contiguous complementary stretches of bases on the same single DNA or RNA strand.

Haploid State The chromosome state in which each chromosome is present only once.

Helix A spiral structure with a repeating pattern described by two simultaneous operations: rotation and translation. It is the natural conformation of many regular biological polymers.

Heteroduplex Double-stranded DNA molecule in which the two strands do not have completely complementary base sequences. Can arise from mutation, recombination, or by annealing DNA single strands in vitro.

Heterozygous Gene Pair The presence of different alleles, for a given gene, on the homologous chromosomes of a diploid organism.

Histones Proteins rich in basic acids (e.g., lysine) found in chromosomes of all eukaryotic cells except fish sperm, where the DNA is specifically complexed with another group of basic proteins, the protamines.

Homologous Chromosomes Chromosomes that pair during meiosis, have the same morphology, and contain genes governing the same characteristics.

Homozygous Gene Pair The presence of identical alleles, for a given gene, on the homologous chromosomes of a diploid organism.

Hormones Chemical substances (often small polypeptides) synthesized in organs of the body. They stimulate functional activity in cells of other tissues and organs.

Host The plant or animal harboring another as a parasite or as an infectious agent.

Hybridization of Nucleic Acid The reannealing of single-stranded nucleic acid chains. The formation of double-stranded regions indicates complementarity of sequence.

Immunoglobulins Y-shaped protein molecules which bind to and neutralize antigen. They are composed of units of four polypeptide chains (2 heavy and 2 light) linked together by disulfide bonds. Each of the chains has a constant and a variable region. They can be divided into five classes, IgG, IgM, IgA, IgD, and IgE, based on their heavy-chain component.

Immunology The study of the phenomena and mechanisms of immunity.

Immunosuppressive Drug Drug that blocks normal immune responses.

Inducers Molecules that cause the production of larger amounts of the enzymes involved in their uptake and metabolism, compared to the amounts found in cells growing in the absence of an inducer.

Infectious Viral Nucleic Acid Purified viral nucleic acid that can infect a host cell and cause the production of progeny viral particles.

Initiation Factors Three proteins (1F1, 1F2, 1F3) required to initiate protein synthesis in *E. coli*.

In Vitro Protein Synthesis The formation of proteins by cell-free extracts.

Karyotype A photograph of a set of chromosomes from a cell, arranged for study in a standardized fashion.

Kinetochore Body which attaches laterally to the chromosomal centromere and is the site of chromosomal tubule attachment.

Ligase (See DNA ligase.)

Light Chains Generally refers to the light protein chains of the immunoglobulins.

Lipids Chloroform or ether soluble organic molecules such as steroids, fatty acids, terpenes, and waxes.

Lysis Bursting of a cell by the destruction of its cell wall or cell membrane, usually by an enzyme.

Lysosomes Internal cellular granules of eukaryotes which contain a large variety of hydrolytic enzymes.

Lysozymes Enzymes that degrade the peptidoglycan in the cell walls of many bacteria.

Lytic Viruses Viruses that cause lysis of the host cell.

Meiosis Germ line diploid cells undergo division to form haploid sex cells.

Messenger RNA (mRNA) The RNA template for protein synthesis.

Metabolism The various chemical reactions occurring in a living cell.

Micron (μ) $1\ \mu = 10^{-6}$ meters or 10^4 Å.

Microtubules Hollow, cylindrical tubules of ~ 250 Å diameter formed by helical aggregation of tubulin molecules.

Missense Mutation A mutation that changes a codon coding for one amino acid to a codon corresponding to another acid.

Mitosis Chromosome duplication and segregation in cell division.

Mitotic Recombination Crossing-over between homologous chromosomes during mitosis, causing segregation of heterozygous alleles.

Monosomy The state in which a cell or an individual has one, rather than two, of any specific chromosome.

Mutation An inheritable change in a DNA sequence.

Nitrogenous Base Aromatic N-containing molecule having basic properties (tendency to acquire an H atom). Important bases in cells are the purines and pyrimidines.

Nonbasic Chromosomal Proteins The acidic nonhistone proteins associated with chromosomes (e.g., DNA polymerase).

Nucleases Enzymes which cleave the phosphodiester bonds of nucleic acid chains.

Nucleic Acid A nucleotide polymer. (See also DNA and RNA.)

Nucleoid The inner core of an RNA tumor virus particle consisting of RNA surrounding by an icosahedral protein shell.

Nucleolus Round, granular structure located in the nucleus of eukaryotic cells. Involved in rRNA synthesis and ribosome formation.

Nucleotide Consists of three components; a nitrogen base (pyrimidine or purine), a sugar (deoxyribose in DNA, ribose in RNA), and a phosphate group.

Operator A chromosomal region which can interact with a specific repressor, and thus controls the rate of transcription of an adjacent operon.

Operon A genetic unit of adjacent genes which function under the control of an operator and a repressor.

Organelle Membrane-bound structure in the eukaryotic cell. It contains enzymes for specialized function. Some organelles, such as mitochondria and chloroplasts, have DNA and can replicate autonomously.

^{32}P A radioactive isotope of phosphorus that emits strong β particles and has a half-life of 14.3 days.

Palindrome A stretch of DNA in which identical (or almost identical) base sequences run in opposite directions, sequence which reads the same forward as backward.

Parthenogenesis Development of an unfertilized egg into an embryo, fetus, or individual.

Partial Denaturation The partial unwinding of the double helix. Those regions that remain intact last are usually GC-rich. G-C base pairs, held together by three hydrogen bonds, are more stable than A–T base pairs (two hydrogen bonds).

Peptide Bond A covalent bond between two amino acids. The α-amino group of one amino acid is bonded to the α-carboxyl group of the other, while H_2O is split out.

Phage (See Bacterial Viruses.)

Phenotype The observable properties of an organism.

Phosphodiester Any molecule that contains the linkage:

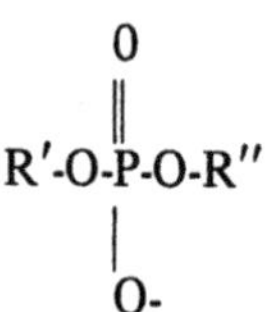

R′ and R″ are carbon radicals (e.g., nucleosides).

Pitch number of base pairs in a turn of the double helix.

Plasma Membrane Physical barrier that surrounds the cell surface and encloses the cytoplasm. Membrane is semipermeable and mainly composed of lipid and protein.

Plasmids Found in bacteria and many other cellular structures. Contain extra-chromosomal circular pieces of double-stranded DNA. These are nonessential to the cell and frequently code for characters such as antibiotic resistance, colicin production and enterotoxin production. They are self-replicating (replicons) as is the bacterial chromosome.

Poly A of Eukaryotic mRNA Relatively long stretches of polyA at the 3′ ends of most eukaryotic mRNAs. They are enzymatically added after transcription.

Polyacrylamide Gel Electrophoresis (PAGE) A process of molecular separation based upon the differential migration of molecules through a polyacrylamide matrix under an electrical potential.

Polynucleotide Linear sequence of nucleotides where the 3′ position of the sugar of one nucleotide is linked through a phosphate group to the 5′ position on the sugar of the adjacent nucleotide.

Polynucleotide Ligase Enzyme that covalently links DNA backbone chains.

Polynucleotide Phosphorylase Bacterial enzyme which catalyzes the polymerization of ribonucleoside diphosphates to free phosphate and RNA.

Polypeptide Polymer of amino acids linked together by peptide bonds.

Polyribosome Complex of messenger-RNA molecule and ribosomes (number depending on size of mRNA); essential intermediate in polypeptide synthesis.

Primary Protein Structure Number of polypeptide chains in a protein and the sequence of amino acids within them. Also the location of inter- and intrachain disulfide bridges.

Primer Structure which serves as a growing point for polymerization.

Prokaryote Simple unicellular organism, such as bacterium or bluegreen alga. Does not have a nuclear membrane or membrane-bound organelles.

Promoter Region on DNA at which RNA polymerase binds and initiates transcription.

Protein A class of compounds composed of a large number of amino acids joined through peptide linkages.

Protozoa Animals composed of a single cell.

Puromycin Antibiotic that inhibits polypeptide synthesis by competing with aminoacyl tRNAs for ribosomal binding site "A."

Recessive An allele which exerts its phenotypic effect only when present in homozygous form, being otherwise masked by the dominant allele.

Recombinant A cell or clone of cells resulting from recombination.

Regulatory Genes Genes whose primary function is to control the rate of synthesis of the products of other genes.

Release Factors Specific proteins involved in the reading of genetic stop signals for protein synthesis.

Renaturation Return of a protein or nucleic acid from a denatured state to its normal configuration.

Repair Synthesis Enzymatic excision and replacement of regions of damaged DNA; e.g., repair of thymine dimers by ultraviolet (UV) irradiation.

Replicon Independently replicating pieces of DNA. The chormosome and plasmids are replicons.

Repressor The product of a regulatory gene, now thought to be a protein and to be capable of combining both with an inducer (or corepressor) and with an operator (or its mRNA product).

Reproductive Engineering Modification of the natural reproductive process; in humans, any generative maneuver other than sexual reproduction, usually involving manipulation of sperm, eggs or embryos.

Restriction Enzymes Components of the restriction-modification cellular defense system against foreign nucleic acids. These enzymes cut double-stranded DNA at specific sequences having twofold symmetry about an axis.

Reverse Transcriptase An enzyme coded by certain RNA viruses which can make complementary single-stranded DNA chains from RNA templates and then convert these DNA chains to a double helix.

ρ **Factor** Protein involved in correct termination of synthesis of RNA molecules.

Ribonuclease An enzyme that cleaves the phosphodiester bonds of RNA.

Ribonucleic Acid (RNA) Nucleic acid occurring in cell cytoplasm and the nucleolus; contains phosphoric acid, ribose, adenine, guanine, cytosine, and uracil.

Ribonucleotide A compound that contains purine or pyrimidine bases bonded to ribose, and is esterified with a phosphate group.

Ribosomal Proteins Group of proteins bound to rRNA by noncovalent bonds to give the ribosome its three-dimensional structure.

Ribosomal RNA (rRNA) The nucleic acid component of ribosomes. It amounts to two-thirds of the mass of the ribosome in *E. coli*, and about half the mass of mammalian ribosomes. rRNA accounts for about 80% of the RNA content of the bacterial cell.

Ribosomes Small circular particles (~ 200 Å in diameter) made up of rRNA and protein; they are the site of protein synthesis.

RNA Polymerase Enzyme that catalyzes the formation of RNA from ribonucleoside triphosphates with DNA as a template.

Rough Endoplasmic Reticulum Large inner membranous sacs (endoplasmic reticulum) which have bound ribosomes. Secretory proteins are synthesized on the membrane-bound ribosomes.

Satellite DNA Eukaryotic DNA which bands at a different density than that of most cellular DNA in equilibrium centrifugation. Often it consists of highly repetitive DNA; it also arises from organelles.

Sex-Linked Genes present on the sex chromosome.

σ **Factor** Subunit of RNA polymerase. It recognizes specific sites on DNA for initiation of RNA synthesis.

Smooth Endoplasmic Reticulum (ER) Extensive inner membranous sacs (endoplasmic reticulum) which are free of ribosomes. The smooth ER may be a major site for attachment of sugar residues to nascent proteins to form glycoproteins.

Spindle Cellular structure, mainly composed of microtubules. Spindles are involved in eukaryotic chromosomal segregation.

Spontaneous Mutations Mutations without "observable" cause.

Sporulation The formation of dry, metabolically inactive bacterial cells with thick surface coats (spores), which can resist extreme environmental conditions.

Streptomycin An antibiotic ($C_{21}H_{39}O_{12}$), isolated from *Streptomyces griseus* (a soil bacterium). It binds specifically to bacterial 30S ribosomal subunits, thus blocking protein biosynthesis.

"Sticky" Ends Complementary single-stranded tails projecting from otherwise double-helical nucleic acid molecules which have poly base ends.

Substrate A molecule whose chemical conversion is catalyzed by an enzyme.

Supercoils Twisted forms taken by covalently closed, circular double-stranded DNA molecules when purification has removed the protein components of the chromosome, thereby slightly changing the pitch of the double helix.

Suppressor Gene A gene that can reverse the phenotypic effect of a variety of mutations in other genes.

Svedberg The unit of sedimentation (S). S is proportional to the rate of sedimentation of a molecule in a given centrifugal field. It is therefore a measure of the molecular weight and the shape of the molecule.

Synthetic Polyribonucleotides RNA made in vitro without a nucleic acid template. Enzymatic or chemical synthesis can be used.

Template The macromolecular mold for the synthesis of another macromolecule.

Tertiary Structure (of a Protein) The three-dimensional folding of the polypeptide chain that characterizes a protein in its natural configuration.

Transcription The enzymatic synthesis of RNA by a process of base pairing, such that the genetic information in DNA determines a complementary sequence of bases in the RNA chain.

Transduction Transfer of bacterial genes from one bacterium to another by a bacteriophage particle.

Transfer RNA (tRNA) Any of at least 20 structurally similar species of RNA, all of which have a MW $\sim$ 25,000. Each species of tRNA molecule combines covalently with a specific amino acid and hydrogen bonds with at least one mRNA nucleotide triplet. Also called adaptor RNA.

Transferases Enzymes catalyzing the exchange of functional groups.

Transformation The process in which a bacterium takes up, integrates, and expresses naked DNA from another organism.

Translation Synthesis of proteins on ribosomes such that the genetic information present in an mRNA molecule controls the order of specific amino acids in the protein.

Translation Control Regulation of gene expression by controlling the rate of translation of a specific mRNA molecule.

Trisomy The state in which a cell of an individual has three, rather than two, of any specific chromosome.

Turnover Number (of an Enzyme) Number of molecules of a substrate transformed per minute by a single enzyme molecule, when the enzyme is working at its maximum rate.

Ultracentrifuge A high-speed centrifuge that can attain speeds up to 60,000 rpm and centrifugal fields up to 500,000 times gravity.

Unwinding Protein Polypeptides that bind to and thus stabilize single-stranded DNA. As a result they tend to unwind the double helix.

Viroids Pathogenic agents believed to consist only of very short RNA molecules.

Viruses Infectious disease-causing agents, smaller than bacteria, which always require intact host cells for replication. Viruses contain either DNA or RNA as their genetic component.

Wild-Type Gene The form of a gene (allele) commonly found in nature.

Wobble Ability of third base in tRNA anticodon (5′ end) to hydrogen bond with any of two or three bases at 3′ end of codon. This enables a single tRNA species to recognize several different codons.

Z-helix Left-handed DNA spiral.

Zygote Union of the male and female sex cells forms the zygote with a diploid number of chromosomes.

Suggested Reading

GENERAL REFERENCES

Office of Technology Assessment, Congressional Board of the 97th Congress. Impacts of Applied Genetics, Micro-Organisms, Plants and Animals. Washington, D.C.: U.S. Government Printing Office, 1981.

BOOK SERIES

Williamson, Robert. Genetic Engineering. Ed. London: Academic Press, 1981.

Setlow, Jane K, Hollaender, Alexander. Eds. Genetic Engineering, Principles and Methods. New York: Plenum Press, 1981.

Streips, UN, Goodgal, SH, Guild, WR. Eds. Genetic Exchange. New York: Marcel Dekker, Inc., 1982.

Wilson, GA. Ed. Genetic and Cellular Technology, Vol. 1, New York: Marcel Dekker, Inc., 1982

Rase, HF. Chemical Reactor Design for Process Plants, Vol. 1: Principles and Techniques. New York: John Wiley & Sons, Inc. 1977

Bailey, JE, Ollis, DF. Biochemical Engineering Fundamentals. New York: Marcel Dekker, Inc., 1988.

BIOCHEMICAL ENGINEERING

Aiba, S, Humphrey, AE, Millis, NF. 2nd ed. Biochemical Engineering, New York: Academic Press, 1973.

Bailey, JE, Ollis, DF. Biochemical Engineering Fundamentals, New York: McGraw-Hill, 1977.

BIOCHEMISTRY

Stryer, Lubert, 2nd ed. Biochemistry. San Francisco: W.H. Freeman and Co., 1981.

Lehninger, AL, 2nd ed. Biochemistry. New York: Worth Publishers, Inc., 1975

Watson, JD, Tooze, J, Kurtz, DT. Recombinant DNA–a Short Course. New York: Scientific American Books, 1983.

BIOLOGY

Keeton, WT, 3rd ed. Biological Science. New York: W.W. Norton & Co., 1980.

GENETICS

Stent, GS, 2nd ed. Molecular Genetics: an Introductory Narrative. San Francisco: W.H. Freeman and Co., 1978.

Watson, JD, 3rd ed. Molecular Biology of the Gene. Menlo Park: W.A. Benjamin, Inc., 1976.

Panopoulos, NJ. Genetic Engineering in the Plant Sciences. Ed. New York: Praeger Publishers, 1981.

Rodrigues, RL, Tait, RC. Recombinant DNA Techniques–an Introduction. London: Addison-Wesley, 1983.

MICROBIOLOGY

Pelczar, Jr., MJ, Reid, RD, Chan, ECS, 4th ed. Microbiology. New York: McGraw-Hill, 1977.

Hawker, LE, Linton, AH, 2nd ed. Microorganisms, function, form and environment. Eds. Baltimore: University Park Press, 1979.

Davis, BD, Dulbecco, R, Eisen, HN, Ginsburg, HS, Wood, Jr., WB, McCarty, M, 2nd ed. Microbiology, including Immunology and Molecular Genetics. Hagerstown: Harper & Row, 1973.

RECOMMENDED JOURNALS

American Chemical Society Winter Symposium*

American Scientist

Chemical Week

---

*Foundations of Biochemical Engineering: Kinetics and Thermodynamics in Biological Systems. Amer Chem Soc Winter Symp 1982.

Genetic Engineering News
Journal of Molecular and Applied Genetics
Nature (British)*
The New Scientist (British)
Plasmids
Science
Scientific American

*Nature 299 (5883): 495-496 (October 7, 1982). A listing of biotechnology texts published in 1982.

Index